On the Gorilla Trail

By MARY HASTINGS BRADLEY

Novels

THE INNOCENT ADVENTURESS
THE FORTIETH DOOR
THE WINE OF ASTONISHMENT
THE PALACE OF DARKENED WINDOWS
THE SPLENDID CHANCE
THE FAVOR OF KINGS

Travel

ON THE GORILLA TRAIL

SISTERS OF THE HUNT

On the Gorilla Trail

MARY HASTINGS BRADLEY

FOREWORD BY MARY ZEISS STANGE

STACKPOLE
BOOKS

Published by
STACKPOLE BOOKS
5067 Ritter Road
Mechanicsburg, PA 17055
www.stackpolebooks.com

Printed in the United States

First edition

10 9 8 7 6 5 4 3 2 1

Cover design by Tracy Patterson
Photos by Carl E. Akeley and Mary Hastings Bradley

Library of Congress Cataloging-in-Publication Data

Bradley, Mary Hastings.
On the gorilla trail / Mary Hastings Bradley ; foreword by Mary Zeiss Stange.—1st ed.
p. cm.—(Sisters of the hunt)
Originally published: New York : Appleton, 1922. With a new foreword.
ISBN 0-8117-3206-1 (alk. paper)
1. Africa, East—Description and travel. 2. Congo (Democratic Republic)—Description and travel. 3. Gorilla—Africa, East. 4. Gorilla—Congo (Democratic Republic) 5. Hunting—Africa, East. 6. Hunting—Congo (Democratic Republic) I. Title. II. Series.
DT361.B6 2005
799.2'7884—dc22

2004056610

FOREWORD

At first blush, the story *On the Gorilla Trail* tells could only strike a twenty-first-century reader as politically incorrect. Its author writes with unalloyed enthusiasm about an expedition with the goal of killing several mountain gorillas. These are the same gentle, endangered primates Dian Fossey studied and brought to world attention in her *Gorillas in the Mist*. Indeed, the hunt that forms the climax of this book took place in the very same region, on the border between present-day Congo and Rwanda, where Fossey lived and died with her beloved silverbacks. And yet here we have a mother who laughs indulgently as her five-year-old daughter plays at "spearing gorillas"[1] and reflects, as she sets out with a female companion for the arduous climb up Mount Mikeno, "I don't wonder that everyone in the country thought the determination of women to see wild gorillas was distinct lunacy. But crawl or climb, Martha Miller and I were going on with it."[2] As this volume's frontispiece makes poignantly clear, the deadly venture was successful.

Mary Hastings Bradley did not herself shoot any gorillas; actually, she does relatively little shooting in this book. She and her husband Herbert Bradley had signed on to this 1921 safari with their friend, the renowned sculptor and taxidermist Carl Akeley, who had long been intent on procuring gorilla specimens for the African Hall he was in the process of redesigning for the American Museum of Natural History in New York City. Like several other celebrated hunting expeditions of the period, this was first and foremost a scientific enterprise.[3] For Akeley, a seasoned African explorer, the expedition would provide not only the dramatis personae for the most important diorama in the Hall, but also a vindication of the life-ways of those massive primates he believed to be our closest evolutionary cousins. If the Bradleys, for whom this was their first African experience, saw it primarily as an occasion for high adventure—a hunting trip par excellence—they nonetheless also grasped that there was some higher or more serious purpose to it as well.

Not all contemporary scholars would agree, of course, as to the legitimacy of this purpose. In her important work, *Primate Visions*, science historian Donna Haraway regards Akeley as a prime exemplar of "the upside down world of Teddy Bear Patriarchy," wherein "in the craft of killing . . . life is constructed." In her in some ways quite astute reading of Akeley's career, she argues that he used his African explorations and the taxidermic installations they yielded—vignettes of stylized, and sometimes subtly anthropomorphized, jungle life—to promote an essentially imperialistic stance toward both nonhuman nature and its nonwhite human inhabitants. Add a generous dash of evolutionary theory, and "natural history" told the story of white male dominance, over nature, over nonwhite men, and over all women. The extent to which women like Mary Hastings Bradley, or like Akeley's two wives Delia and later Mary Jobe, bought into this story testified to the power of the patriarchal myth of white manhood, especially among those classes of females who stood to benefit from association with powerful white men.[4]

There is some obvious merit to this line of reasoning. European exploration of the "Dark Continent" went hand in hand with its imperialist exploitation, which was carried out with a steadfast belief in the natural (indeed, God-given) superiority of whites over nonwhites, and of men over women. Human superiority over animals and plants almost went without saying. In the mindset of patriarchal colonizers, hunting went along with militarism; they were parallel forms of conquest and subjugation.[5]

During the period of colonial expansion, European women tended as men did to accept as fact this "pseudo-scientific racism," which, historian Patricia Romero points out, "lingered among some colonials as long as they were part of the ruling class."[6] Given the history of institutionalized racism in this country, it is hardly surprising that an American like Bradley would entertain similar views, nor that she would be as unaware of her own racial prejudices as she often seems to be in the narrative that follows. She remarks, for example, on similarities between Capetown, South Africa, and her native Chicago:

> The mingling of black with white upon the streets was a note of foreignness to the English visiting South Africa for the first time; but it had a certain homelike air to us, and it was a distinct shock to hear some starched mammy burst into Kaffir to upbraid her youngsters.[7]

To the extent that "blacks" (native Africans) and "coloreds" (most of whom were Indians) had marginally more rights in Cape Province at this time than in the rest of British-ruled South Africa, the comparison is probably apt. Still, it is hard not to gasp when, a few pages later, she likens Cecil Rhodes's place in South African history to Abraham Lincoln's in our own.[8] And she reveals a disturbingly blithe indifference to the racist implications of the following parallel she draws between the "boys" who provide the labor on safari and the workforce of the pre-Civil War South:

> Our relation to these boys in those months of safari was very much, I imagine, the relation that existed before the Civil War between kindly owners and their black possessions. These boys were at our service for any one of the twenty-four hours, for just as much service as we could get out of them. But they were amply safeguarded from overwork . . . by their native leisureliness and lack of responsibility.[9]

There is, then, clear evidence in Bradley's narrative to support Haraway's indictment of the Social Darwinism underlying early-twentieth-century exploration.

Yet this book deserves a fairer and more careful reading than Haraway may have given it.[10] It is virtually impossible to read hunting literature from the early twentieth century without encountering the kind of casual, and often far less unselfconscious, biases one finds in Bradley's writing. Books always speak from, as well as to, their times. But during the time in which Bradley was hunting and writing about it, public perceptions of hunting and conservation were changing, and Carl Akeley's work was a significant factor in spurring this change. In this time before television and widespread tourism, most people learned most of what they knew about wildlife and geography from books, zoos, and natural history museums. Working for several of the latter, Akeley developed a technique and a philosophy that established taxidermy as the best way to portray animals in their wild state. (If it seems ironic that dead museum specimens were thus preferable to live zoo animals, it must be borne in mind that animals in zoos of the period rarely had the opportunity to "act naturally.") This meant that scrupulous on-site observation of the animals in their natural habitat was as essential as procuring exemplary individuals to mount. The goal was to convey to the public as accurate and lifelike a depiction of wild animal life as possible, thereby raising general consciousness about conservation issues.[11]

When it came to gorillas, achieving this goal presented an unusually tough challenge. As Bradley points out below, up until the time of their safari there was "no gorilla in a museum mounted by a man who ever saw a wild gorilla."[12] Well into the twentieth century, this largest of primates was more a figure of myth than of natural history. Popular ideas about these reclusive animals had been shaped largely by the sensationalist writings of a Franco-American explorer and self-styled scientist named Paul Du Chaillu, who in the mid-nineteenth century wrote two book-length accounts of the lowland gorillas he had encountered in western Africa. In Du Chaillu's rendering (some of which is quoted by Bradley[13]), the gorilla was a "hellish dream creature," with a volcanic temper and a penchant for attacking terrified humans (particularly females of our species, it would seem). Du Chaillu made a name for himself on the lecture circuit, alternately enthralling and appalling his audiences with descriptions of the perils he, the fearless White Hunter, faced in gorilla country: The animal was "an impossible piece of hideousness. . . . One blow, with its bony paw, and the poor hunter's entrails are torn out, his breast bone broken or his skull crushed."[14]

Despite the fact that Du Chaillu's native porters disputed some details of their "Bwana's" account, and that a subsequent explorer, the American Winwood Reade, reported that gorillas were actually prone to attack only in self-defense and then usually in warning, the image of the gorilla as prototypical King Kong was still the predominating one in Akeley's time. Akeley aimed to change this, not only because he had read between the lines of earlier accounts and suspected gorilla behavior to be far different than had been reported, but also because he was convinced that the gorilla was closely related to human beings.[15]

In the early twentieth century, a German explorer named von Beringe had discovered gorillas in Uganda, and scientists quickly decided that these mountain gorillas were a distinct and independently evolved species. These great apes were to be Akeley's quarry, and he dedicated himself to their study with an unusual leap of faith, composing in longhand a "gorilla creed" to guide his work. It reads in part:

> I believe that the gorilla is normally a perfectly amiable and decent creature. I believe that if he attacks man it is because he is being attacked or thinks he is being attacked. I believe that he will fight in self-defense and probably in defense of his family; that he will keep away from a fight until he is frightened or driven to it.

Akeley went on to say he also believed that the "man who will allow a gorilla to get within ten feet of him without shooting is a plain darn fool," given the gorilla's superior size and strength. Noting that his "faith in the general amiability and decency of the gorilla" was based on deductions from the observation of other animals in the wild—primarily monkeys and lions, both of which species fight largely in defense of self and offspring—he turned to further reflections, based upon a gorilla he had observed many years before at the London Zoological Park:

> It was very young and its chief aim in life seemed a desire to be loved. This has seemed to be the chief characteristic of the few gorillas that I have seen in captivity. They appear to have an extremely affectionate disposition and to be passionately fond of the person most closely associated with them.

After recounting the story of a circus gorilla who was thought to have "died of a broken heart because he was separated from his mistress," Akeley concluded, "The above notes are here set down for the purpose of recording the frame of mind with which I am going into the Kivu country to study, photograph and collect gorillas."[16]

Enter Mary Hastings Bradley. She and her husband Herbert, Akeley's attorney as well as his friend, had long entertained the idea of joining Akeley on one of his specimen-collecting trips. So when the invitation came early in 1921, they, and perhaps especially Mary, were delighted to sign on. She was by that time an established novelist and journalist and apparently eager to follow in the footsteps of those late-nineteenth-century "traveling ladies," women of generally substantial means who had undertaken (sometimes as companions to their husbands or fathers, more rarely on their own) adventures hitherto unthinkable for members of the frailer sex. Mary Bradley was among the generation who wrote about their explorations between 1900 and 1930. These women, according to journalist and geographer Elizabeth Fagg Olds, were largely American and were different from their trailblazing predecessors in that they were less inclined to be "romantic dreamers," driven by a taste for the exotic, and more likely to be goal-oriented realists. They constituted, she suggests, "a transitional group in the evolving advances of women, for they were the direct forerunners of today's trained women scientists and field workers."[17]

While Bradley took a less active interest in field work than did contemporaries of hers like Courtney Borden, or Delia or Mary Akeley, she fits the profile in that she was an independent, professional woman in her own right. Born and raised in Chicago, she had graduated from Smith College. In 1919, two years before the gorilla expedition and nine years after her marriage to Herbert, the *Smith College Weekly* published an editorial proclaiming: "We cannot believe it is in the fixed nature of things that a woman must choose between a home and her work, when a man may have both. There must be a way out and it is the problem of our generation to find the way."[18] Bradley sought that way. When the African opportunity arose, one of her first decisions was that her five-year-old daughter Alice would go, too. To those who objected that the equatorial jungle, filled as it was with beasts and savages, was no place for a child, she retorted that "Alice was as safe in Africa as in Chicago. Safety means ceaseless vigilance in either case."[19]

Upon their return from Africa, *The New York Times* reported that "the jungle has no fear for Mrs. Bradley; she says she feels safer there than on Broadway."[20] This report may have been a bit of an exaggeration, since in fact Bradley confesses to a goodly amount of fear in the narrative which follows, as when she wryly remarks, "In the Congo your worst fears are never realized. Something that you didn't fear happens instead."[21] In this regard, too, she is like her sister-explorers, who as a group tended to be more prone frankly to acknowledge their fears than male adventurers, whose primary objective was generally to emphasize their own heroic exploits in the most dramatic light possible.[22] Akeley appreciated this quality of Mary Bradley in particular, and of women in general. In February of 1921, he wrote the Bradleys:

> The object of our expedition is gorillas and gorilla country—the last word in Africa adventure. I cannot possibly come back from this expedition as a hero because I am taking ladies and a little baby into a country that is full of beauty and charm, and I shall not be able to tell a tale of hardships and dangers overcome; but it is just as much fun to pull down the pedestals of fake heroes as it is to build the pedestal for yourself.[23]

There was nothing self-effacing about Carl Akeley, but he did appreciate good press when he saw it. He knew that one excellent way to challenge the prevailing image of the gorilla as a bloodthirsty brute, with, it was said, a particular proclivity for kidnapping human women, would be to bring several women into gorilla country. At the same time, he was angered when the Bradleys themselves pitched their adventure story to the press. He regarded this as sensationalizing at the expense of "Museum dignity," and he was particularly annoyed that they told reporters they were going to Africa to "hunt" gorillas. "Say that I am to study gorillas," he wrote them in July, very shortly before they set off on their journey. "One of the big objects of the whole expedition is to kill the idea that hunting gorillas is to be considered a sport. Everyone is getting mad to hunt gorillas and I hope that I will be able to influence legislation to prevent the hunting of gorillas except for strictly scientific purposes."[24]

This letter suggests, as does his "gorilla creed," that for Akeley there was a larger legacy at stake, and for a hard-living man approaching sixty and not in the best of health, this collecting expedition was something of a pilgrimage. Akeley's biographer, Penelope Bodry-Sanders, believes he underwent a "*metanoia*," a complete and life-changing conversion while enveloped in the mists of Mount Mikeno. He looked into the face of "the Old Man of Mikeno," a gorilla he had killed (and would later mount for the African Hall), and saw himself. After that expedition, he took to signing letters, "The Old Man of Mikeno." His empathy for the gorillas was such that he felt impelled to lay down his gun in favor of the camera. Indeed, he said he was glad Herbert Bradley was there to shoot the big silverback who would dominate the diorama as the "Lone Male of Karisimbi," because he, Akeley, could not have brought himself to have killed him.[25]

Herbert Bradley, of course, was only too happy to have done it, since shooting a gorilla was what he went to Africa for in the first place. His wife's feelings appear to have been rather more mixed. Her reflections on the death of the big male have an elegiac quality: He was "a huge, shaggy, primordial thing, like something summoned out of the vanished ages. And the scene in which he lay had a beauty that was like nothing earthly." Contemplating his face in death, she describes his "mild and patriarchal dignity," his apparent "prescience that something was happening in the world against which his strength was of no avail—as if he knew the security of his high place was gone."[26]

If in the gorillas of the Virunga mountains Akeley had seen his own humanity, Bradley saw in them the personification of an Africa vanishing in the face of advancing civilization, a theme (hardly original with her) that forms a leitmotif in this book. Like a number of other writers of the period, Bradley simultaneously expresses a sense of wonder and gratitude that she has been able to experience these natural treasures firsthand and a sense of urgency regarding the need to aggressively protect endangered

Africa. She stresses how the expedition actively practiced conservation; with ten permits in hand, they killed only five animals, and those were for the museum. She argues, very likely at Akeley's behest, for the removal of the gorilla from game lists and the creation of a preserve. That was actually to be accomplished over the next several years. In 1923, the Belgian King Albert decreed the creation of the Parc National Albert in the Virunga region. By 1929, the boundaries of the half-million-acre park had expanded into present-day Rwanda (to include the area where Fossey would work). It was the first genuine wildlife sanctuary in Africa.[27]

Bradley concludes her narrative of the gorilla expedition with words of gratitude:

> It had been worth it all; the heart-breaking climb, the cold, the discomforts were merely incidents, a price one gladly paid—especially in retrospect!—for the rare experience of seeing gorillas in their savage solitudes.[28]

But this is barely halfway through the book. The Bradleys had come to Africa in search of adventure and plenty of that lies ahead in the remainder of the story, including several riveting encounters with lions, interactions with a variety of natives and colonialists, and an abundance of "local color," particularly relating to the inner dynamics and politics of a safari on the move. It all adds up to a valuable historical document. It is also a very satisfying read.

On the Gorilla Trail would be the first of three hunting books authored by Mary Hastings Bradley. She, Herbert, and Alice returned to Africa for a second safari in 1924, and she recounted their further adventures in *Cannibals and Caravans* (New York: Appleton and Co., 1922). In 1928, she and Herbert headed to Ceylon (modern-day Sri Lanka) and India, and then on to Sumatra, to hunt tigers, a journey Bradley retells in *Trailing the Tiger* (New York: Appleton and Co., 1929). She also worked with little Alice to produce two children's books, for which her daughter provided the illustrations: *Alice in Jungleland* (New York: Appleton and Co., 1927) and *Alice in Elephantland* (New York: Appleton and Co., 1929). After that, with the exception of some time she spent as a war correspondent in Europe during World War II, Bradley settled down in Chicago. Turning her attention to her hometown, Bradley published a four-volume novelized history of the Windy City under the title *Old Chicago*. In addition, she wrote historical romances and murder mysteries and garnered two O. Henry Awards for short fiction in 1931 and 1938. Her last novel appeared in 1952. Mary Hastings Bradley died in 1976.

Curly-topped Alice Bradley, who was at many points the "start attraction" in the story recounted in *On the Gorilla Trail*, would eventually follow in her mother's footsteps as a writer. But this was only after a stint in the U.S. Army, where she served in intelligence and achieved the rank of major, participated post-discharge in the creation of the Central Intelligence Agency, worked as an undercover agent in the Near East, and then enrolled at American University in her early forties to earn a B.A., and in her early fifties earned a Ph.D. in Experimental Psychology from George Washington University. In the late 1960s she began publishing highly-regarded science fiction under the name James Tiptree, Jr. She also wrote fiction under the name Racoona Sheldon (Sheldon was her actual married surname). Her identity as Tiptree came to light after her mother's death in 1976, the same year in which Tiptree—lauded by critics for "his" ability to draw strongly convincing female characters—won one of three Nebula Awards. Alice Bradley Sheldon died in 1987. Since 1991, the Tiptree Award has annually been given, in her memory, to the work of science fiction or fantasy that best expands and explores gender roles.

Carl Akeley made one more trip to Africa, returning to the Varunga region—the place, he liked to say, "where the fairies dance"—with his second wife, Mary Jobe Akeley. Ailing with fever, he made the arduous ascent up Mount Mikeno. The party set up camp with a view of Mount Karisimbi, near the spot where Herbert Bradley had shot his silverback. Akeley died there on November 17, 1926, five years to the day after the "Lone Male" fell. His wife Mary completed the expedition and oversaw the completion of the African Hall. The American Museum of Natural History's Akeley Hall of African Mammals opened in 1936 and remains, as Akeley designed it, open to the public today.

Mary Zeiss Stange
Ekalaka, Montana
July 2004

NOTES

1. See below, 80.
2. Below, 81.
3. See two other volumes in the Sisters of the Hunt series for examples of other such specimen-collecting hunts: Osa Johnson's *Four Years in Paradise*, recounting an African adventure in which Akeley and the American Museum of Natural History also figure; and Courtney Borden's *The Cruise of the Northern Light*, which tells the story of an Arctic voyage sponsored by Chicago's Field Museum.

4. See Donna Haraway, *Primate Visions: Gender, Race, and Nature in the World of Modern Science* (New York: Routledge, 1989), Chapter 3, "Teddy Bear Patriarchy Taxidermy in the Garden of Eden, New York City, 1908–1936."
5. On the history of European expansionism in Africa, see Thomas Pakenham, *The Scramble for Africa: The White Man's Conquest of the Dark Continent from 1876 to 1912* (New York: Random House, 1991). On the role played by hunting in imperialism, see especially John M. MacKenzie, *The Empire of Nature: Hunting, Conservation and British Imperialism* (Manchester and New York: Manchester University Press, 1988), and also Valerie Pakenham, *Out In the Noonday Sun: Edwardians in the Tropics* (New York: Random House, 1985), Chapter 7.
6. Patricia W, Romero, Editor, *Women's Voices on Africa: A Century of Travel Writings* (New York & Princeton: Markus Wiener Publishing, 1992), 20. On the ready acceptance of the natural superiority of the ruler over the ruled, see also the foreword to Agnes Herbert and A Shikári, *Two Dianas in Alaska*, another volume in this series.
7. Below, 26.
8. Below, 33.
9. Below, 166. A display in the restored slave quarters at South Carolina's Rose Hill Plantation State Park describes the attitudes of benevolent slaveholders toward their easygoing and lethargic slaves, who were naturally suited to working long hours in hot sun, in strikingly similar fashion. Rose Hill was the home of secessionist governor William H. Gist.
10. Haraway lists *On the Gorilla Trail* in her bibliography. However, when she describes in detail the image of the slain "Big Gorilla of Karisimbi," which appears as this book's frontispiece, she makes clear that she is describing a photograph in the American Museum film archive, and in an endnote she complains that "Reserving it for internal use only, the Museum refused permission to publish this photograph. Is it still so sensitive after 68 years?" (*Primate Visions*, 34, 386 n. 18) Since *On the Gorilla Trail* was in the public domain by the time she was writing, she could certainly have copied the frontispiece, had she been aware of it. Considering the weight this particular image carries in her discussion, it is odd that she would have forgotten seeing it in Bradley's book. Nor, in her discussion of the safari this book recounts, does she cite *On the Gorilla Trail*; all of her references are to Carl and Mary Akeley's writings.
11. Not surprisingly, Akeley was one of the first to appreciate the possibilities of live-action photography, as both a research tool and a way to entertain and educate the public about wild animals. He invented the Akeley Camera,

which he used himself and also gave to Martin Johnson, whose work for the American Museum of Natural History he arranged.

12. Below, 4.
13. Below, 105.
14. Quoted in Penelope Bodry-Sanders, *Carl Akeley: Africa's Collector, Africa's Savior* (New York: Paragon House, 1991), 182.
15. He was only slightly mistaken there, of course. While we are indeed members of the same primate family, our nearest relatives are chimpanzees.
16. The entire "gorilla creed" is quoted by Bodry-Sanders, *Carl Akeley*, 183–184.
17. Elizabeth Fagg Olds, *Women of the Four Winds* (Boston: Houghton Mifflin Company, 1985), 2.
18. Quoted in Nancy F. Cott, "The Modern Woman of the 1920s, American Style," in Francoise Thébaud, Editor, *A History of Women: Toward a Cultural Identity in the Twentieth Century* (Cambridge, MA: The Belknap Press of Harvard University Press, 1994), 83.
19. From Bradley's *Cannibals and Caravans*, quoted by Kenneth Czech, *With Rifle and Petticoat: Women as Big Game Hunters, 1880–1940* (Lanham and New York: The Derrydale Press, 2002), 129.
20. *The New York Times* April 2, 1922, p. 56. ProQuest Historical Newspapers, *The New York Times*.
21. Below, 147.
22. Romero, *Women's Voices on Africa*, 10. My own reading corroborates this insight of hers.
23. Letter dated February 3, 1921. Quoted in Bodry-Sanders, *Carl Akeley*, 177.
24. Letter dated July 11, 1921. Quoted in Bodry-Sanders, *Carl Akeley*, 178.
25. See Bodry-Sanders, *Carl Akeley*, 194–196.
26. See below, 115-117.
27. For a good contemporary narrative of the evolution of the Parc National Albert, and the various actors in addition to Akeley who helped make it happen, see Mary Jobe Akeley, *Carl Akeley's Africa* (New York: Blue Ribbon Books, 1929), chapter XIX, 239–255. Today renamed Virunga National Park, along with Rwanda's Volcano National Park and Uganda's Mgahinga Gorilla National Park, it was in 1979 named a World Heritage Site by UNESCO. More recently, in 1994, it was declared a World Heritage Site in Danger. The 1.4-million-acre Virunga mountain region is endangered by various forces: war in Congo and Rwanda, the influx of Rwandan refugees into Virunga Park, deforestation, and poaching.
28. Below, 135.

THE BIG GORILLA OF KARISIMBI

CONTENTS

CHAPTER | PAGE

I. Off to Africa 1
II. The West Coast Trip 11
III. The Tourist Trail 25
IV. In the Belgian Congo 38
V. On Safari 57
VI. The Summit of Africa 71
VII. The Gorilla Trail 98
VIII. The Big Gorilla of Karisimbi 110
IX. A Gorilla Band 119
X. The Pygmies Come to Camp 136
XI. A Night in a Crater 145
XII. The Three-Stone Kitchen 157
XIII. The Lion That Came to Life 168
XIV. Lion Hunting at Night 185
XV. Elephants and Buffalo 199
XVI. Santa in the Jungle 212
XVII. Across Uganda 220
XVIII. The Tomb of King Mutesa 233
XIX. Good-by to Africa 247
XX. Lists and Equipment, Etc. 257

ILLUSTRATIONS

Facing Page

The Big Gorilla of Karisimbi . . . . *Frontispiece*
The *Kenilworth Castle* at Madeira 12
Freetown, Sierra Leone 12
General Smuts 13
Victoria Falls—Main Falls from Rock, beyond Livingstone Island 34
Mrs. Bradley at Victoria Falls, The Devil's Cataract . 34
Where Do We Go from Here? Alice Hastings Bradley on the Congo 35
Steamer on the Lualaba River 35
Borassus Palms on the Lualaba River 44
Congo Belles Traveling 44
Crocodile with Arm and Leg of Native He Had Just Eaten 45
Gavial, Asiatic Species of Crocodile, Caught in Lualaba River 45
Giant Palm on Tanganyika 50
Native Boat on Lake Tanganyika 51
The Tree at Ujiji beneath which Stanley met Livingstone 51
Native Market, Usumbura, Lake Tanganyika . . . 54
Native Barmaids at Usumbura 54
Porters on the March 55
Alice on Safari 55
Personal Attendants on the March—Camera, Gun and Wheel 60
Alice and Mablanga 60
Alice and Her Tent 61
Porters Waiting Rations 61
Rest House and Chief's House in the Rusisi Mountains . 68
Martha Miller and Her Elephant 68

ILLUSTRATIONS

Facing Page

Wife of Erstwhile Cannibal Chief, Kabaka 69
A Chief of the Warundi, Lake Tanganyika 69
Watussi Family Bringing Milk and Butter 90
Watussi Girls Making Butter by Shaking Cream in a Gourd 90
White Father's Mission at Nyunde 91
Cathedral in Erection at Nyunde 91
White Sisters at Nyunde 94
Mt. Chaninagongo 94
Native Fishing in Kivu 95
Flow of Lava across Lake Kivu 95
The Woman with a Hoe 100
Our Postman 100
The White Fathers at Lulenga Mission 101
Lulenga Valley. Mission White Fathers 101
Our Objective—the Gorilla Triangle 112
Gorilla Camp on Mt. Mikeno 112
Where Gorillas Live—the Fairy Forests of Karisimbi . 113
The Big Gorilla of Karisimbi Shot by Herbert E. Bradley 113
Mr. Akeley Working on Gorilla Skins 120
Guides Removing Flesh from Gorilla Skeletons . . . 120
Gorilla Skeletons—and Others 121
Gorilla Camp—Skeletons and Skins Drying—Little Clarence Hanging in Tent 121
In the Gorilla Forests of Mt. Mikeno 124
Gorilla Bed and Trail 124
Gorilla Bed Overhung by Ferns 125
Just before Meeting Gorillas—Mrs. Bradley and Porters 125
The Peak of Mt. Mikeno, 14,600 Feet 130
Porters' Huts at the Gorilla Camp 130
A Glade in the Bamboos 131
Native Cattle in the Foothills of Mikeno 131
Batwa Chief and Wife 140
Dance of the Batwa 140
Crossing the Lava Plains 141
A Night on the Summit 141

ILLUSTRATIONS

Facing Page

Lava in Eruption in Nyamlagira's Crater 154
The Fire Pot of Nyamlagira—Taken at Night by Its Own Light 154
Sunny Jim 155
Our Bicycle Boys 155
Leo and the Union Suit 166
The Three-Stone Kitchen and Its Aids 166
The Ruchuru River 167
Killed by Lion—Grave of Mr. Foster 167
Mrs. Bradley and the Lion That Did Not Stay Dead . 182
The "Dead" Lion That Roared as His Picture Was Taken 182
Topi Staked Out for Lion Bait 183
Natives on the Ruindi Plains 183
One Night's Kill—Mr. Bradley with Two Lions Shot by Him 196
Miss Miller and Her Lion 196
Natives Making Fire with Fire Sticks 197
Ready for the March 197
The Acacia Is Fantastically Flat Topped 222
Buying Bark Cloth beneath the Euphorbia or Candelabra Tree 222
The Golden Crested Kavirondo Crane 223
Family Scene at Lake Bunyoni 223
Waiting for Canoes on Lake Bunyoni 226
Native Dugouts on Bunyoni 226
The Tomb of King Mutesa, in Uganda 227
The End of Safari 227
Ripon Falls—the Birth of the Nile 250
Alice at a Ceremonial Dance of the Kikuyus 250
Natives Sawing in Nairobi 251
A Rickshaw at Mombasa 251
Giant Baobab Tree at Mombasa 254
Good-by to Africa 254
Memories 255

ON THE GORILLA TRAIL

CHAPTER I

OFF TO AFRICA

An Expedition for Gorillas Which Includes a Five-Year-Old Explorer

Our objective was a tiny triangle in the heart of Africa. It was bounded by three volcanic mountains and was a high plateau of bamboo forest, eternally cold, eternally clouded, eternally rainy.

It was there that Mr. Akeley was going to find gorillas for a group for the American Museum of Natural History of New York, and we were going with Mr. Akeley because we wanted to see Africa, as well as gorillas, and the way to this triangle was through one of the loveliest and least known parts of the continent, the Eastern Congo. No Americans had yet been in the country to which we were going.

I had always wanted to see Africa. I suppose I first thought about it when, like all Sunday-school children, I shouted,

> Where Afric's sunny fountains
> Roll down their golden sand,

and wondered if the life of a missionary did not have its thrilling compensations in its intimacies with crocodiles and cannibals.

ON THE GORILLA TRAIL

I know that Africa first touched my imagination when my great-grandfather read aloud to me his favorite book, Stanley's *In Darkest Africa.* I received then a vivid intimation of Africa's mysterious spell, stirring pictures of a vast continent peopled with savages, of feverish jungles and mighty rivers, of treacherous beauty and swift death, of a primitive barbarism that had been going on from the beginning of time, unchanged and unchanging, living its own life through the centuries, unknown and untouched by trade or civilization.

I made up my young mind then that I would go and see Africa and that resolution was kept alive by our family friendship with Mr. Carl E. Akeley, then with the Field Museum of Chicago. Mr. Akeley had already made one expedition to Africa and later he and Mrs. Akeley went on two expeditions, and from that time we saw Africa through the Akeleys' eyes.

The Dark Continent was transformed. It was Africa the Beautiful, a land of wonder and delight, of wide plains and mighty forests and glacier-peaked mountains, a world of tropic splendors roamed by primitive peoples and magnificent beasts. It was Mr. Akeley's enthusiasm which inspired Colonel Roosevelt to make his African trip, recorded in *African Game Trails,* and that same year Mr. John McCutcheon of the Chicago *Tribune* was with the Akeleys on a hunting trip in British East, where he wrote *In Africa.*

I used to stand before the Fighting Bulls in the Field Museum, a pair of elephants shot by Mr. and Mrs. Akeley, a group which is a record of his sculptural methods of mounting, and wonder if I should ever be able to see an elephant on its native heath—and then live to remember it. So, with all this interest in the subject, my husband and I were delighted that the gorilla expedition came at a time when we could arrange to go.

The delight was not unanimous. In general our announcement was received by our friends with a flattering gloom, with what might be called the bedside manner to those resolved upon an untimely end. The less solemn and concerned were frankly facetious.

Why do it this way, they wanted to know. Why not the lake or chloroform or something usual and immediate instead of taking the trouble to go to Africa and give ourselves to a lion for lunch? Also, demanded the merry ones, if anything *did* happen, were we planning to ship the consumer back to the Zoo so they could place memorial wreaths about its neck?

They called attention to Mr. McCutcheon's picture of the jolly little cemetery back of Nairobi with "Killed by Lion" on every cross, and quoted "The bulge was Algy," with persistent humor. Especially they pointed out that the gorilla was not noted for hospitality and presented us with various accounts in which we invariably came upon some such heartening paragraph as, "The poor brave fellow who had gone off alone was lying on the ground in a pool of his own blood, his entrails torn out, his gun beside him, bitten in two by the gorilla's teeth."

I admit there was room for both humor and dismay. Between the lion's chances for lunch and ours for a rug the odds were sportingly even. As to the gorilla, the records were not encouraging, describing the grown male as a demon of ferocity, attacking on sight with a fury few hunters can withstand; but the records of the gorilla were extremely scarce.

It is surprising to learn how little has been discovered about the animal since Du Chaillu wrote his blood-curdling accounts of his adventures in the West Coast jungles in 1848. A few recent pamphlets, a few isolated instances, comprise the world's authentic information. There is no gorilla in a museum mounted by a man who ever saw a wild gorilla. Skins have been bought from hunters and collectors and stuffed according to the best available information. Almost nothing of the animal's habits or capacities has been discovered. As far as we could find out, only four true gorillas ever reached the United States alive, three short-lived youngsters and, in 1920, the famous John Daniels, who did not long survive the separation from the English woman who had brought him up.

Mr. Akeley had no intention of bringing back a gorilla alive—although for some moments he dallied with the idea and I held an agonized breath, seeing myself walking the floor with the wailing infant—but he wished to study the animal as much as possible, to bring back the material for a group, to make anatomical and scientific records of every kind, and, if it were possible, a photographic record, something that had never before been done. He was going equipped with his own invention, the Akeley Motion Picture Camera, to try to realize that dream.

Personally Mr. Akeley believed that the rumors of

the animal's unreasoning ferocity were exaggerated. He believed that the male of the species was maligned and that unless he was attacked or his family threatened his intentions were honorable and unobtrusive, and that he was a harmless and interesting old gentleman who ought to be taken off the game lists.

And Mr. Akeley proposed to find out. He proposed, after he had secured his museum group, to give the gorilla every social opportunity to meet him halfway.

My husband and I were unbiased. We were neither for nor against the gorilla. He might be as peaceful as unexploded dynamite. The reputation of the species might be due entirely to the irascibility of minority hotheads. We were not going to take any position beforehand, but let the gorilla show himself in his true colors. Privately I believed that my best position later would be behind something substantial.

I was sustained by that word *wary* which I found in every account of the gorilla. Wary and elusive were his invariably given characteristics. Now I rather liked that in him. He could be as elusive as Peter Pan. I had no intention of frustrating any social barriers he wished erected. My New England blood could be as proudly reserved as his. He could rely upon me not to make undue advances.

I was going into his country. I was trying to penetrate his domain and spy upon affairs that were undoubtedly his own concern, but I was not going to thrust myself and what might be an uncongenial New World personality upon his attention. And while I had no intention of killing a gorilla I had no intention, either, of strolling through his impenetrable bamboos without a gun. There was always the possibility that one might be dealing with the leader of the minority hot-heads!

But the fact that we were going to gorilla land was no serious cause for concern. The difficulties of discovering gorilla were so great that we might feel ourselves fortunate if we got a glimpse of them at all. The real concern was in the fact that we were going to Africa, as a friend expressed it, with a gun in one hand and a baby in the other.

Mr. Bradley and I were taking our five-year-old daughter Alice with us. But it wasn't as mad as it sounded. Mr. Akeley's experience would not burden the expedition with a child unless it were both safe and feasible. We were going into a healthy region, up from the Cape through the Belgian Congo to Lake Kivu, along the mountainous backbone of Africa, where, although almost under the equator, the altitude would insure cool nights and pleasant days.

Alice was an outdoor child who loved the open, and the experience would be an unforgettable part of her life. Our camp would be comfortable and well protected, and her chief danger would be the equatorial sun against which a helmet and unceasing vigilance could guard her. Eternal vigilance is the price of existence at home even, where motors menace every curb and crossing, and where she could never for an instant be safely left alone upon the streets.

It seemed to us that the same unceasing care which tried to guard Alice in Chicago could keep her safe in Africa and that the change and outdoor life would be a tremendous benefit to her. We decided it only after careful study of every book of travel and unending consultation with every one we knew who had been to Africa; then we made up our minds to go ahead and meet every moment with thorough care but not to invite apprehension. So our party, headed by the most experienced of African travelers, included the youngest of explorers.

Once the decision was made, the matter of outfit was upon us. Tents, camp equipment, and "chop" boxes of food were ordered by Mr. Akeley from London firms to be ready for shipment when we arrived; the various cameras—motion picture, plate and film—and the developing apparatus were all collected here by Mr. Akeley, and the guns were arranged for here. My husband and I each had rebuilt Springfield rifles, 30-30, with hard and soft nose ammunition, and in addition my husband had the gun with which Mr. McCutcheon had slain his elephant, a .475 Jeffery.

Ordering by catalogue is an enticing joy, but after that was done we were left to struggle with the thousand and one details of personal things for the long sea voyage as well as for the interior. My lists ranged from hobnailed shoes and flannel shirts and khaki knickers to white crêpe and lace evening gowns. Only a woman who has tried to estimate the hairpins she will need for months in the wilds, and how many pairs of stockings and how many boxes of colored crayons a lively little girl will use, can feel for my state of mind!

We outfitted in Chicago during a spell of July weather that made us wonder weakly why on earth we were trying to get any nearer the equator anyway; we packed, with the valiant assistance of friends, complicated trunks for hold and baggage room and stateroom —with the inevitable after-panic lest the gold slippers be in the hold and hunting trousers appear in the stateroom!—we bade farewells that savored almost of the eternal and turned our backs upon home and family and friends and the familiar perils of civilization.

We set sail upon the *Baltic* for Liverpool on July 30, 1921. Besides Mr. Akeley and the three Bradleys the party included Miss Martha Akeley Miller, Mr. Akeley's secretary, and Miss Priscilla Hall of Chicago who was to be Alice's special guardian. Mr. and Mrs. Leonard Baldwin of New York accompanied us as far as London.

We had a crossing whose restful calm and six meals a day fortified us for the next few hectic days in London, pursuing Last Things. These included helmets and spine pads for the sun, and Jaeger blankets and heavy pajamas for the cold mountain nights, and mosquito boots for camp wear, and air-tight cases for packing, and a formidable medicine kit; and our rooms at the hotel looked a collector's paradise.

The Very Last Thing was a can presented by a thoughtful English officer, received with hilarity and later acknowledged with reverent gratefulness—a little can of insecticide.

London seemed little changed from the London of before war days, but there were two very poignant reminders of the war—the Cenotaph in Whitehall to the Glorious Dead, and the grave of the Unknown Warrior in Westminster Abbey. The service was just over the morning I went to the Abbey, and the dim aisles were filled with a throng that made its slow and quiet way up to the high paling about the grave. Within the palings I saw a stone slab covered with wreaths and little nosegays. Later there would be a permanent slab of Belgian marble and a dedication in brass lettering, but now in the dark stone was the simple inscription:

AN UNKNOWN BRITISH SOLDIER, KILLED IN WAR,

FIGHTING FOR COUNTRY AND FOR KING

Greater love hath no man than this.

The wreaths of all nations were there, with imposing names, but the most touching things were those little nosegays of violets and pansies, tied sometimes with a string. As I stood there I saw a little bunch of field flowers put through the palings from the pressing crowd and I looked down to see a young girl with country cheeks and eyes as blue as the cornflowers she had brought. She was crying.

"It might be her brother, you know," said the Englishman beside me.

It was clear from the wet eyes of most of the women and the faces of the men that they were at the grave of son or father, brother or husband. The world moves quickly and all too easily forgets, but these were the people who would never forget, and this was their Place of Pilgrimage.

For Alice, London consisted of hansom cabs and the Zoo. Her ambition to ride up beside a cabby, behind a real horse, was gratified by the discovery of an old-fashioned four-wheeler at Charing Cross and Alice was lifted up to ride in triumph through the interesting streets. All the motors in the world were nothing to her beside the delight of that. She and the cabby, as old-fashioned as his vehicle, agreed perfectly.

" 'Orses be the thing," said he. "Gentlefolk should 'old by them." He had a very poor opinion of machinery indeed. "It's the undoing of the 'uman race," he told us.

At the Zoo she rode a donkey, a camel, an elephant, and a llama in preparation for strange African mounts.

CHAPTER II

THE WEST COAST TRIP

Sierra Leone and the Burning Saxon; an Impression of General Smuts

The West Coast of Africa has an ominous ring. It savors of jungles and fever and hot stagnant harbors to which the sickly caravans wind down. . . . Gold Coast and Ivory Coast. . . .

Probably no one knew less about the West Coast boats than I did, and when we found that the West Coast passage was the only one we could reserve in advance from New York I made inquiries that later filled me with amusement. Assured that the trip was comfortable and not at all fatal we took heart and passage.

We sailed from Southampton on the *Kenilworth Castle,* Friday, August 12, and our first discovery was that we were not the only family in the world risking its young upon the West Coast. The decks were full of children and almost all, we learned, had been born in Africa or had gone out at a tender age and were now returning from leave at Home.

It looked as if all the African cradles had not been robbed by lions or emptied by fever. It also looked—and proved—extremely social for young Alice. If not the only child, she was the only little American, and as such received a special mothering from Ruth Smith, a sweet little South African girl.

The *Kenilworth Castle* proved a most comfortable boat, the only ominous sign being the information that electric fans could be rented from the Barber Shop. But we never needed them. We had a delightful trip, a succession of lovely Junelike days, never too warm even when crossing the equator.

We went direct to Capetown, a seventeen-day trip, with one official stop at Madeira the fifth day out. Four days at sea is enough to make land an event. We anchored before dawn and came early on deck—a deck we found festooned with embroideries and invested with swarthy venders—to look across blue waves to Madeira, a picturesque mountain island, smothered in green, with here and there a gleam of cream brick villas and red roofs. We went ashore in small boats and up steep stone steps to the quay.

Our destination was a hotel halfway up the mountain and we had a choice of vehicles—the native sledge, drawn by oxen or donkeys over the tightly packed little stones with which the narrow streets are paved, or a motor car.

Hunger prevailed over picturesqueness, and we motored through the tiny town with a blaring horn scattering beggars and urchins, and climbed the steep mountain to Reid's Palace Hotel where we breakfasted over a paradise of a garden, with shining views of sea and sky. The breakfast itself deserves honorable mention and was our introduction to the delicious passion fruit—a fruit the size of a lemon with a hard

THE KENILWORTH CASTLE AT MADEIRA

[page 12]

FREETOWN, SIERRA LEONE

[page 12]

General Smuts

[page 19]

rind and a soft, rosy-purple interior, and it was served on old mahogany that made one begin to compute freight rates.

After a walk through one of the loveliest gardens imaginable—palms and tropic bloom and high walls dripping with flowering climbers and everywhere an outlook over the blue bay far below—we went down into the little tourist-trap of a town and compromised upon embroidered handkerchiefs, mindful of African luggage problems ahead.

It was that same day that the Committee on Sports met. Now that might not seem a momentous thing to record—until you knew that committee. Inspired by one dynamic spirit it arranged field sports, cricket matches, bridge drives, chess games, and tournaments of tennis, shuffleboard, deck quoits and bucket quoits, ladies' singles and doubles, gentlemen's singles and doubles, and mixed doubles. So well did it do its work that no one ever sat in peace thereafter, with a weak inclination towards a book or a pipe, without being reminded that he must now play off his tennis with Major Miller or that Captain Viney was ready for the finals in deck quoits.

The day after Madeira we crossed the Tropic of Cancer and wore hats on the upper, unroofed deck, and people began to talk about the sun and I began to wonder how Alice would ever remember in Africa about a helmet on her curly head.

I heard the encouraging little anecdote about the friend of Mr. Powell's who had taken off his helmet to wave a friend good-by and was dead in two hours in consequence, and other anecdotes of even swifter disaster, and I fortified myself with the idea of hiring a lynx-eyed native whose sole duty would be to stalk after Alice and prevent catastrophe.

Wireless brought us the news that the *Saxon,* which had left Southampton the week before us, was on fire and putting back on her course. We were to meet her at Sierra Leone and take off as many passengers as possible. So we saw more of the West Coast than the palm-fringed strip of Cape Verde which we passed early on the nineteenth, for on the twentieth we steamed past a flat coast line, straight on, apparently into a mountain, then round a green promontory into a lovely bay, Freetown harbor, where the *Saxon* lay at anchor, smoke still pouring from her blackened portholes.

Freetown was on the mountain side, the barracks on the top. Most of the population of Freetown seemed to be in canoes or square-rigged sailboats streaming out to us with limes and breadfruit and baskets.

It was sweet to see how high sympathy ran for the unhappy Saxonites. We offered to double, to treble, in our staterooms that they might be accommodated, and we heard with scorn the stories which trickled from the purser's quizzically cynical lips of those who, single in staterooms, had approached him with tales of coughs or snores which would make life impossible for any roommate—although, of course, personally, each one would have been only too glad, etc., etc. . . .

But that night, looking sympathetically off towards the *Saxon's* lights, against the dark shadows of Freetown, and listening to the endless creak of the cranes swinging the luggage aboard, we discovered that into the rooms which we ourselves were trebling up to vacate, only two Saxonites were being put.

It is sad to record how human nature changed. Why were not the Saxonites three in a room? Was it, perchance, because they were the daughters of a general? The spirit which had laid the Magna Charta before King John approached the purser with this injustice, which he acknowledged with the swiftness of one quickened with experience, and altered his arithmetic.

But as a matter of record by doubling and trebling we took on almost all first-class passengers and we never heard the slightest murmuring against any possible compression, and, as it proved, we were infinitely indebted to the *Saxon's* disaster, for the passengers we gained from her added greatly to the interest and pleasure of the trip.

They were a distinguished line as they came on board next morning—Sir Arthur Lawley, former governor of the Transvaal, with Lady Lawley and their daughter; Sir Lionel Phillips of South African Gold King fame; the Duc d'Orleans, who, if France had a king, would be that king, a big, blue-eyed, blonde-bearded man going out with his physician for big game hunting in British East; and, conspicuous among them, the tall, military figure of General Smuts returning from the alleged Peace Conference with Ireland.

My first and instant impression was of the soldier in Smuts, the air of authority, of responsibility, of quick and stern decision. I saw a strong, dignified face of guarded reserve, blue eyes with the keen glance of a scout, bushy brows, gray hair, a closely trimmed yellow mustache, the cropped suggestion of an imperial, an aquiline nose, and firmly molded mouth and chin.

It was the face of a man who has fought and fought hard and is unwearied. It had the restraint and thoughtfulness and indomitable tenacity of statesmanship, and in everything about him was the soundness and vigor of a splendid physique in the prime of power.

We did not go ashore that morning, warned of the difficulty of the return in small boats against a seventy-mile-an-hour tide. If we had gone we should have seen the inhabitants wending to their eighty-six churches—some churches have congregations of only three and four—clad frequently in white night shirts and silk hats. I yearned for the sight.

Freetown was founded some eighty years ago by the slaves returned from slave ships by English gunboats. It has a population of forty-one thousand blacks of conglomerate races and sixty whites. There is a black London barrister, and a black editor who is a B.A. of Cambridge. It could not have been that editor who wrote the advertisement which I saw in a Freetown paper, the *Sierra Leone Echo and Law Chronicle* of August 20, 1921:

BUNGIE

> The General sympathetic undertaker, builder for the living and the dead, contractor, etc.

Refuge and refreshing bungalow, 15 Kissy Street, Freetown.

Ready-made coffin supplied with hearse and uniformed men at any moment, corpse washed and dressed. Ready-made hammock always in stock for sale.

Trucks, Venetian blinds, etc., all repairs and payment after satisfaction.

And I'll bury your dead by easy system, only be honest to your sympathetic last friend.

Bungie's advice: Do not live like a fool and die like a big fool. Eat and drink, pay your honest debts—that's Gentleman, always praying for a happy death then a coffin by Bungie. Will bury the dead Book of Tobias I'll feed the living, that's Bungie.

Contracts taken for Carpentry, Masonry, Painting Tombstones, etc., at moderate charges.

R. LUMPKIN *alias* ALIMANY BUNGIE.

Our sympathetic friend had evidently joined no union and was no specialist.

From an old English resident of Freetown I gleaned the information that native wives cost five shillings, ten for a "starched one," and in the event of no offspring after a year the wife is returned to her father who refunds the shillings. It seemed to us that the father might have been allowed a slight deduction for rent or depreciation, but no, he had to refund the price entire.

In all our preparation for the trip, in all our mountainous equipment, it had never occurred to us to plan for a Fancy Dress Ball. Only the experienced English came provided. But for the Barber Shop we should have been lost. The Barber Shop supplied everything for just such unknowing ones. Miss Hall, however, would have none of the offerings and with burnt cork and a gunny sack she carried off the humorous prize as Topsy, much to General Smuts' delight.

The children had as social a time as we, with their own field sports and games and their masquerade. Alice went as a French doll in a box supplied by the obliging ship's carpenter and, to her frank joy, won a prize. For her birthday, which fell on the twenty-fourth, she had a party and a marvelous cake, arranged for by shipboard friends, the Days; and we, who had feared rather a lonely voyage for her, now looked forward to the conclusion of the excitement of the trip.

We crossed the Line the morning of the twenty-third. There were no ceremonies. It was not hot and the night after a girl was wearing a fur coat between dances, although the deck was screened with bunting. Not a day of the eighteen was uncomfortable; though lacking a breeze, they might have been warm. For a character firm enough to eschew deck sports it would be an ideally restful trip.

Our last celebration was the second concert, followed by the distribution of prizes to the athletic ones and an ebullition of speeches. General Smuts held forth so glowingly on the high moral rewards of living the simple life in South Africa that we were almost persuaded to leave the effete civilizations and come out and live it.

A whisper saved us. "The simple life of a Prime Minister," said a South Africander in our ears.

But it was a good speech. It had the force, the humor, the magnetism and inspiring enthusiasm which showed us General Smuts in action. The shipboard days had already given us vivid pictures of him. He was genuinely interested in the serious scientific possibilities of the expedition and a little amused and appalled at women and a child venturing into the wilds. He called us his "dear gorillas."

"At the end of the Boer War," he said, one afternoon, "we had irregular fighting—what the English call guerilla warfare. But the Dutch did not understand that word. 'The damned English called us gorillas,' they said," and the general laughed heartily in reminiscence. There was good-natured banter but not a touch of rancor in him. Wise Smuts, they call him.

There are no dull phrases in his conversation. He cuts to the heart of a subject. His questions are incisive and direct, his speech is vigorous, animated, shot with humor. He asked interestedly of America, of the problems there, of prohibition, immigration, and the women of America, their activities and their home life.

"To succeed, a nation must have fine women," he said, "big, splendid women," and you saw in him the strength of his hearty Dutch blood and the pride of strong, self-reliant ancestry.

"One looks to America," he said earnestly, and that

brought us to the Peace Conference and President Wilson.

"He came as God," he said. "The people of Europe were hungry for good, for the things of the spirit. You understand? That was the thing he was to them. It was the secret of his enormous prestige. But no man could do it. It was beyond human power—the passions of men that had to be reconciled. I was there. My wife and children were in Africa, and for six months I fought the terms of the peace treaty as hard as I could fight. I saw much of Wilson and House. I know the whole story. But it was too much for man to do. Only God could do it. I said that one night at a meeting. I said, 'Now is the time for the Griqua prayer.' "

He explained that the Griqua is a mixed race, some Hottentot, a very little white. But they are Christians. "Now there was to be a battle between the Griqua and the blacks the next day and the Griqua came to God in prayer. You ought to hear it in broken Dutch. But it was like this. He told God he had often prayed before and been disappointed. God had failed him. Now to-morrow was to be the great battle. 'Blood will flow,' said the Griqua. 'It will be a terrible thing. Now, God, you be there. Come yourself. Don't send your Son. This is no place for children. Come yourself.'

"I told the Peace Conference that night, 'This is the time for the Griqua prayer!' "

He laughed, his eyes twinkling. His laugh, his humor, is a great reason for his success. You can see him winning over his opponents, heartening his tired soldiers, joking with bluff Dutch farmers.

"Humor is the saving of us," he declared. "It is the salvation of our race."

Of Wilson he said again, "He was not God, and no one but God could have done it—not a mere erring mortal like ourselves."

I said that I thought in America the time had come for the Griqua prayer, and the talk went back to American problems again.

"I am reading *Main Street,*" said General Smuts, and he asked about the truth of the picture it drew. He found the doctor in the story magnificent, operating away with the tools at hand, at night, on a farmhouse table; while the wife, who escaped from it all, did nothing after she got away—nothing but talk.

"Main Street!" he said, humorously. "All my life I have been a Main Streeter!" and he chuckled. "All Main Streeters, we fellows who are trying to get things done—trying to do something besides talk about it—working away with the things at hand; and the other fellows, who would make such a different world of it if they were God, criticizing and tearing away! If they were God—they would make a fearful mess of it!" he flung out with a flash of sternness.

Several times he spoke of his convictions of the separate origins of life.

"Unquestionably it has had separate origins," he said. "You understand——?"

He had a way of saying, "You understand?" or "Do you follow?" with an intent look from those searching eyes of his as if he were saying, "If you are at sea, speak out. Life is too short to talk incomprehensibly."

He went on, "It has sprung up and died out and sprung up in other places. And we may go—just as the Neanderthal man has gone. We may go. The problem of life is too much for man. We are in this frame of earth and God has given us a soul . . . and we strive and fight . . . and the consciousness of the world and the sorrows of it wear us out."

He stopped. "The only happy man I know is the black. He is a distinct race. The black will work all day, work as hard as you can make him. But night comes, he eats his bellyful, he sings. He has the secret of happiness."

He touched his breast, half smiling. "We others, we have too much here. It is too much for us, and we may go and another race take our place."

It struck me as characteristic of the man that he should have this feeling so strongly, should accept this with scientific detachment as a possible conclusion to all our human endeavors and yet be, in his infinitesimal span of life, not at all detached, but one of the hardest workers to achieve results that no hope of his could call permanent.

There was nothing tragic, nothing frustrate, in his face. He has the steady courage of the man who had looked life and death in the eyes and marched through defeat to continued effort.

He is a man who believes in the old substantial foundations—country, home, and family. He believes in hard work, in enthusiasm, in endeavor, in good cheer.

His roots go deep into the soil, into the good, strong, warming earth. Well educated—it was Cambridge, I think, that he went to—it is his native endowment of sound, penetrative good sense, disciplined in a hard school, that is his unfailing inspiration.

A South Africander gave me a very vivid picture of the General's wife.

"She's independent," he told me. "Not a bit proud. But independent. I mean, she'd do anything she liked about her house—she'd do the washing if she thought she wanted to and if the King of England should come to call she'd not be a bit put out, but make him welcome and go right on. She's got eight children. There's usually been a baby in one arm and a book in the other. She's a great reader of very serious books."

I don't know how true a picture that is—I didn't ask the General—but I thought it a delightful one.

The youngest of those eight children is a little girl about Alice's age, and for Alice and her care in the interior the General gave me a great deal of good advice. At the very moment of disembarking, I remember, he came back from some official group with a last word of recommendation about her.

The night of that last concert was what might be termed blowy. It reminded us that we were approaching the Cape and that the Cape was first called the Cabo Tormentoso, or Cape of Storms, by its sorrowful discoverer Dias as his mutineers drove him back past it in 1486.

Rough weather was generously predicted. And the next morning seemed to us rough. The ship's officer denied it. It was too heavy a sea for a life boat to be launched, and a strong swimmer's agony could not possibly last two minutes—but it wasn't rough. It wasn't officially rough until the fiddles were on the table. Now our fiddles were not on. The ship had an end-for-end plunge with a shuddering whirl and list which kept the soup swirling round and round so no racks were needed. But there were many aboard who did not know that it was not rough. Some one should have told them.

CHAPTER III

THE TOURIST TRAIL

CAPETOWN, BULAWAYO, THE GRAVE OF CECIL RHODES, AND VICTORIA FALLS

CAPETOWN has a magnificent approach. It ranks with Rio de Janeiro, Naples, and San Francisco as one of the most beautifully situated seaports of the world. I have never seen Rio de Janeiro, but I can imagine no more stirring entry than to sail through the Cape's blue waters into mountain-guarded Table Bay, flung round with the peaked Apostle Range but dominated by the dark, precipicelike height of Table Mountain. The town lies at its feet, between the rocky outlines of the Lion's Head and the Devil's Peak.

The white tablecloth of cloud was rolling off the mountain's level top as we came in; flags were flying in honor of General Smuts' arrival, reception committees stood waiting, mounted police in line. There was a festive air to the disembarking.

Leaving the men to struggle with the luggage and interview the courteous customs officials, we four feminines drove to our rooms at the Grand Hotel for which we had wirelessed. The Mount Nelson is generally frequented by tourists, but it was a little too far away for our multitudinous errands.

It was the end of August, still Capetown's winter, a day of lovely sun outdoors, but within the hotel we found that coolness and chill so reminiscent to Americans of the Continent and Cathedrals. We viewed with joy the announcement that there was a fireplace in every room, but the joy was short-lived. There had never been a fire in any fireplace. There never was.

But we spent little time indoors during the days we waited for our express train north. The Capetown streets are vivacious, many of the shopping ones galleried from sun and rain, with open balconies for tea—which is not only an afternoon function but an eleven o'clock one as well—and with windows hung with tempting ivories—of which we mistakenly imagined we were to see more as we neared the source of the ivory supply.

Flowers were everywhere. Twice a week the Flower Market masses the spring bloom before the post office, the most striking to us being the great bunches of calla lilies which grew wild; every day the venders offered a luxuriance of violets and narcissus. The trim little gardens of the trim little brick villas in the town were solid sheets of golden daffodils, and in the beautiful residences which made up the suburbs the flowers and vines seemed to us a tropical profusion, although we were told that at the end of the cold season the flowers were only beginning.

The mingling of black with white upon the streets was a note of foreignness to the English visiting South Africa for the first time; but it had a certain homelike air to us, and it was a distinct shock to hear some starched mammy burst into Kaffir to upbraid her youngsters.

The population is almost evenly divided as to color. The 1914 census gave the Europeans as 80,863 and the colored as 74,555.

The town is a metropolis, with an air of energy and enterprise; but the place I liked best in it was the Avenue leading out of Adderly Street, an old Dutch Avenue, laid out by Governor Van der Stel, and shaded by gnarled oaks of almost two centuries.

The Avenue reaches for three-quarters of a mile, bordered by the Gardens and the Museums and the foliage-smothered Government Houses, so that the city itself does not encroach upon it. No vehicle is allowed to enter, and a saunter there in the late afternoon, with the Lion's Head glowing ruddy above the oaks, has a tranquil charm of atmosphere all its own.

There is history in Capetown. The Cape itself is a romantic name, reminiscent of old Portuguese adventurings, and the history of Cape Colony is the story of Dutch enterprise and the East India Company, of burghers in exile and gallant merchantmen, of wrecked treasure and small-pox and Kaffir raids and English wars. The Cape Colony numbered nearly twenty-seven thousand when, in 1815, it became a permanent English possession and its story became an English story of advancing borders, Kaffir raids, and Boer wars.

The glory of Capetown is in its drives. There are marvelous roads, cut in the rocky coast line of the Cape, winding in and out among the mountains for miles and miles. The only thing comparable to them is the Corniche Drive, and even Riviera memories paled before the splendor of these wild and lonely scenes—the mountains rising on one side and the bloom and glitter of the sea below.

I can think of no drive more beautiful than the one past Sea Point and Camp's Bay, edging the Twelve Apostles' range to Hout's Bay, then crossing the Peninsula to Constantia and Wynberg and back to Capetown through the Rhodes Estate. The day we went all the trees were in leaf except the oaks; calla lilies were blooming by the wayside; the mountain sides were rosy with heather; the sea was half a hundred blues. At Constantia Bay there was a lovely stretch of silver trees against the jade green of the Indian Ocean.

Those trees, whose pointed leaves were like velvety silver, are peculiar to the Cape. The southern and eastern slopes of Signal Hill are clothed with them, while they can hardly be induced to grow on the other side of the mountain.

Some day, before so very long, there will be a swift passenger service from New York to Capetown and those South African roads will be filled with a swift succession of motorists. It will be a paradise for the first-comers.

Outside of Capetown is Groote Schur, an old-fashioned white-washed residence of Dutch architecture, which Cecil Rhodes set aside in his will as the official residence of the Prime Minister of that South African confederation he foresaw. It is a quaint irony of history —which not even Rhodes could have foreseen—that the first two men to occupy it should have been of that people he so dreaded—General Louis Botha and General Jan Smuts.

THE TOURIST TRAIL

From Capetown to the Congo runs the southern part of the road that was Rhodes' dream. Draw a line up from the Cape, along the very backbone of Africa, to a point almost under the equator, and close to the Eastern frontier of the Belgian Congo, and you have the way that we were going. Our objective, the three volcanic mountains—Mikeno, Karisimbi, Visoke—lay northeast of Lake Kivu in what was formerly German East Africa, now part of the Belgian Mandate. Rail and water would take us north and after that our feet.

We left Capetown Saturday, September 3, in a comfortable compartment train, with a good diner attached. We rented bedding and towels from the black boy who appeared each night to make up our beds. The train went twenty miles an hour and the comfort and quiet and cleanliness of it made the trip very much less tiring than one a quarter of the length in America, while the strangeness of the scenery kept us always at the windows or on the little seats on the platform's end.

All the first day we climbed through the Hex river valley among the mountains which form an apparently insurmountable barrier to the Great Karroo tablelands above. The next morning we looked out upon the wide expanse of veldt, grass and brush and flat-topped granite kopjes—a few herds grazing, a few lone farms, a few stray ostrich, but no game—a graveyard of plains emptied of its old life by the Boer farmers who have not replaced it with cultivation.

Dust was beginning to sift in and the cold morning had warmed to sultriness. We were exhausting the resources of the compartment wash basins when a note was brought in from the private car at the end of our train. Sir Arthur and Lady Lawley, who had been on the boat with us, had learned of our presence on the same train and Lady Lawley wrote to offer us, not the tea alone which my mind foresaw at the first sentence, but a nap for Alice in her bed and baths for us all in her tub.

We restrained ourselves. We accepted the bath for Alice only and at a convenient stop at Orange River that afternoon Miss Miller and I raced back with her. It was the longest train I ever saw, bar none, not even the freights which roll past when motoring—and then, confronted by the Lawley tub in the flesh, we succumbed and bathed in grateful turn and "as welcome as Lady Lawley's tub" became a familiar phrase with us.

The end of the second day brought us to Kimberley where the diamonds come from. Kimberley is a snapshot of impressions—an outline of diamond mines against the paling sky—wide streets, low buildings, and shady pepper trees—the hospitality of *Kenilworth Castle* acquaintances; a white tropic house and charming court; hasty motoring and farewells.

In 1869 you could not find Kimberley on any map. In 1870 a shining little stone was picked up by a Boer farmer. In 1871 ten thousand men were digging frantically in that precious earth—sixteen hundred claims, each thirty-one feet square, were pegged out.

Access to these claims was difficult, for the narrow strips required to be left for roadways soon became mere bridges as the ground was excavated; water filled in, slides occurred, dust fell from one claim to another, quarrels and shooting affrays led to law-suits and the diggers became impatient of the situation and resentful of increased taxes and rents.

Exasperated or discouraged, they sold their claims, sold them to one or the other of two shabbily dressed men who happened invariably to turn up—and Cecil Rhodes and Barney Barnato had the diamond industry of South Africa in their hands.

Their amalgamation, now known as the De Beers Consolidated Mining Company, is one of the greatest coups in the history of finance. The first thing they did was to close the greater part of the mines so that a limited output would preserve the high prices.

The name of Mafeking got me out next morning, at what I then considered an unearthly hour of six-thirty, to see a long station of stalls, one huge pepper tree, two aloes, three palms in tubs, and four natives selling dusters. I strolled out and discovered the outskirts of the town, hot and empty streets, one-story stores, flagged sidewalks, when sidewalks there were, and came back to the stalls to imbibe bad coffee in a cracked cup from a cross-eyed proprietress.

I felt an increased respect for that gallant band who had defended the town so successfully. Personally I should have surrendered it at the first opportunity, or even paid them to take it away. . . . But then the coffee was exceptionally bad.

Little black boys in increasing décolleté began to beguile our stops with offerings of grotesquely carved giraffes—made in Japan!—and a few native huts appeared. We rushed out and photographed them as if they were the last huts we should see in Africa. In the brush and veldt I often caught a likeness to our western sagebrush country, but just when the scene appeared familiar some swift touch changed it utterly—a huge cone of a sun-baked ant hill, a scattering of little crooked trees, or a blaze of scarlet flowers on bare branches.

The morning of the fourth day brought us to Bulawayo where we motored thirty miles across the veldt through Matapos Park to World's View.

In Bulawayo there stands Rhodes' statue—a stocky, thick-set, carelessly dressed man, hands behind him, legs apart, looking out to the north where his dream road stretched. The body of that man lies among the granite boulders on a mountain top, beneath a simply marked bronze slab.

Forceful, hard-headed, dreamer and laborer, far-seeing, indomitable, and unsparing, his name has become identified now with that Empire for which he wrought with such passion of will in spite of her. It is one with that dream of an All Red railroad from Cape to Cairo, of which the world has heard so much that a good part of the world takes for granted the dream's accomplishment.

He is not alone upon his mountain top. Down at the beginning of the climb there is cut in the rock,

THIS PLACE IS CONSECRATED
AND SET APART FOREVER
TO BE THE RESTING PLACE OF THOSE WHO
HAVE DESERVED WELL OF THEIR COUNTRY

and a little way below Rhodes' grave, at one side, is a white marble memorial guarding the bodies and memory of Allan Wilson and his thirty-three men who fell against the Matabele—gallant soldiers perishing for that Empire for which the master spirit on the rock beside them wrought so unflinchingly.

From the top the mountainsides sloped down into forest and brush, far-reaching until they met the hills on the horizon. As we were there, a troop of baboons came into the trees, their strange barking emphasizing the solitudes.

It is a wonderful resting place, rock and height and encircling view of his loved land, and if his unresting spirit could lie quiet anywhere it is there in the place he had designated. He went unready and untired to his end—his life work, as he thought, unfinished, but his soul has gone marching on, like John Brown's.

"Keeping faith with Cecil Rhodes" will be the politician's slogan for years, just as the name of Lincoln is with us.

At Bulawayo we changed trains and discovered that an unfeeling power had removed the diner. We had spent at Rhodes' grave the time perhaps intended to be employed in collecting food, but at a small station farther on the kindly wife of the station-master parted with some of her own supply for us and half her precious bread for our little girl. We lunched and dined upon these gleanings, and it might be said that we had a distinct interest in the breakfast at the hotel at Victoria Falls the next morning.

The very sight of the hotel was charming, a long, one-story building with raying wings. From the verandas at the back we could see the gorges of the Falls and the bridge which Rhodes had planned, which was completed in 1904. The thunder of the Falls was part of the place like the rhythm of heavy surf on the sea coast.

The many views of the Falls themselves are at some distance and the way is covered by little trolleys, tiny cars for ten or twelve, on miniature tracks, operated by man-power. The man-power, a variable number of natives, seizes the car and pushes. It is simplicity itself, except perhaps for the natives.

From the landing stage to which the trolley brought us we took a native canoe out to Cataract Island. There are two large islands—Cataract and Livingstone—perched on the very brink of the Falls, and the Devil's Cataract is that plunge of water between Cataract Island and the South Bank.

It was an unforgettable first glimpse, a mad leap of water over savage rocks, into a deep and narrow gorge, with a barbaric blaze of red aloes on the banks against the white foam. The color and beauty of the country were all a part of the picture. It was utterly wild and lonely, as untouched and solitary as when Livingstone in 1855 first solved the mystery of the Smoke that Thunders—Mosi-oa-Tunya—the native name for the Falls, born of the white overhanging cloud of vapor that can be seen for twenty miles.

There is no comparison between Niagara and Victoria. To say that Victoria has two and a half times the height, four times the volume of Niagara, and a width that is the reach of ten city blocks, tells nothing.

Mrs. Bradley at Victoria Falls,
the Devil's Cataract
[page 34]

Victoria Falls—Main Falls from
Rocx beyond Livingstone Island
[page 34]

WHERE DO WE GO FROM HERE?
ALICE HASTINGS BRADLEY ON THE CONGO

[page 42]

STEAMER ON THE LUALABA RIVER

[page 42]

Niagara is a spectacle, an unchallenged spectacle, for it has the distance and perspective from which it can be seen—Victoria is a sensation.

The Zambezi, the second largest river in Africa, comes rolling quietly out of the very heart of the continent, and in a flash its whole width plunges blindly down into the deep and narrow chasm through which it rushes, raging and storming, for more than forty miles. The wild and savage splendor of it makes Niagara a benign performance. The rocks on the brink of the plunge shatter the racing water into diamond points. It is spray when it begins to fall, foam and fury and fairy crystals, outflung like powder from some giant gun, hurtling into air to run together in sheer-dropping columns, that fall a shuddering depth into thundering gorges below, to rise as spray again shot with the gleam of double rainbows.

At Victoria we donned our gray pith helmets, companions of many days to come, and my fears for Alice were laid. No lynx-eyed native would be needed to stalk after her and prevent carelessness. She never forgot, she never objected, and she reminded the rest of us with fairly irritating consistency.

On the eve of departure we left a call for five in the morning. The train to the Congo was due at six-twenty and an hour and twenty minutes was not too much time. Now while the dining room of the hotel was excellent and the main rooms were charming, the bedroom service had a certain negligence of attention from the management, betokening a mind preoccupied with other things.

So we were insistent that our call should be loud and punctual.

The management was amused. It said it never failed.

I woke at six. A tropic dawn was lightening the world. My watch dial showed six. In twenty minutes the train would roll out of the station. In the moment, after the leap from the bed and the reach for the nearest garment, came a knock at the door.

"Six o'clock," said the management's representative.

We knew it. What we said there is no use to record. Mr. Akeley and Mr. Bradley were in the hall, shouting for porters for our trunks and bags and dressing between shouts. I sped into the nearest things and while Miss Hall jammed down the suitcase tops I drew a coat over Alice's pajamas, thrust on sandals and socks, topped her with a helmet and in seventeen minutes, breathless but triumphant, we stood in front of the station, followed by galloping porters with luggage, and the rest of the party on their trail.

We stood there an hour and twenty minutes. We saw day dawn brighter and brighter, revealing our unwashed faces and sketchily done hair. The grass and flowers took on the color of light, the dogs came down to investigate our boxes, and play with Alice who, with pajamas flapping beneath her coat, scrambled about the mound of luggage.

The station was yet unopened and the station clock had stopped. I saw a black boy, sent by the curio dealer across the tracks, come with a clock in his hand, regard the station time-piece intently, and then, as he had been told to set his clock by the station clock, he put the hands conscientiously at a quarter after four and ambled back again.

Ultimately the train came, and a day and a night and another day of it brought us to Sakania, the Belgian frontier.

CHAPTER IV

IN THE BELGIAN CONGO

FROM ELIZABETHVILLE DOWN THE LUALABA RIVER TO LAKE TANGANYIKA

WE entered the Belgian Congo on September 10, at ten-thirty at night, and my first impressions were of darkness and confusion, of a long, low building with wanly glimmering lights, returning travelers bustling about, porters swarming round little fires on the earth by the tracks, and dogs shifting from group to group in the peculiarly apprehensive manner of native dogs.

Here we presented passports, arranged for the luggage to go through in bond to Elizabethville, and changed from the British railways to the Katanga railways of the Congo.

A night and more than half another day brought us to Elizabethville, the capital of Katanga province, where we were met by representatives of the absent vice-governor and motored to the hotel where rooms had been arranged.

We had come now twenty-five hundred miles of the northward way and we knew one thing already about the equatorial gorilla. He was inaccessible. No casual sportsman, landing on the coast, was going to corner him.

We waited a week in Elizabethville, because the luggage which had left Capetown with us, and for which we were paying luggage rates and not freight in order to have it accompany us, had disappeared. We had seen it last at Livingstone, just this side of Victoria Falls, and persistent telephoning disclosed that it was still there. We dislodged it at last and it arrived in a cattle car, too late for the first connection north, for only two trains a week run to the interior from Elizabethville.

The Rhodesian Railway, I may add, saw no reason whatever in the cattle car journey to rebate anything of the rate we were paying to insure the retention of that luggage in a luggage van. They are still refraining from seeing it, although we are continuing the correspondence in a public-minded spirit of enlightenment.

Elizabethville is young and the early residents still talk of their first tin houses, but it is built now for the future, well laid out, with wide streets and attractive administrative buildings. There are clubs and tennis and golf and motors—in spite of the dollar and a quarter for gasoline—good shops and Belgian and English libraries. The copper mines have brought quite a few Americans there, and our wait was enlivened by everything that the new acquaintances and the courteous acting vice-governor, Monsieur Serruys, could do for us.

We were very comfortable at the Hotel de Bruxelles, our first characteristic tropic hotel, with its stretch of veranda before the dining room, reading room and bar, through which one indiscriminately sauntered to reach the bedrooms beyond, at right angles to the main building on either side of a court.

Madame Franzos, the proprietess, was a presiding genius, or rather an ambulatory one, and to see her march majestically up and down the court was to see a general on scout duty. Her shout of "Boy" meant violent retribution.

We were getting acquainted now with the boy system. If we wanted boots or a bath we put our heads out a window or door and shouted "Boy!" into the resounding court. There is a directness about this lacking to the buzzer system. If you don't get your boy you can always raise your voice, but you cannot always raise your buzzer. Its drawbacks are for the other guests who may be wanting a nap when you want your boy, but no system is universally considerate.

We got some boys ourselves in Elizabethville. We had hoped to find there a complete safari outfit for the interior, gun boys, tent boys, cook and interpreter such as Mr. Akeley had always obtained in British East, but there was nothing of the sort to be found. The Congo was not invaded by travelers and we would have to pick up what material we could as we went along.

We found three house boys whom we took with us —three with whom we were to spend many months of experience. Little they knew as they stepped off that Sunday morning in the glory of the private car which the hospitable Belgian government furnished us, again through the courteous offices of Monsieur Serruys, what fate was holding in reserve for them.

Mablanga, Kiani, and Merrick they were, new-shod,

clean trousered and shirted, check-capped, with Merrick sporting an artificial red rose upon his purple shirt. The rose was all that survived of his outfit. It had many transient backgrounds, but the most humorous of all was the suit of pink cotton pajamas which Miss Miller supplied at a Congo camp to relieve his very evident needs.

We paid each of these boys the scandalously high wages of fifty francs a month—three dollars and fifty cents as the rate of exchange then stood—and thirty-five cents a week to provide food. Most Belgians, we discovered later, paid but half that and I am afraid that our trio had the bright idea of taking advantage of our inexperience. But considering all that fell to their lot while they were with us I cannot feel that they were overpaid.

The country through which we went was increasingly beautiful—a rolling country of wide valleys and splendid slopes, overflung with forest like a Persian rug. There was a soft, rusty-rose tone of spring leaves that wove in and out like an elusive pattern; there were deeper notes of red from hanging tree fruits; there were gleams of gold and orange and crimson from flowering shrubs and trees, all interwoven in a tapestry of woodland greens and browns, soft and subdued, changeful with sunlight and cloud shadow.

But for all the glory of color the charm of African landscape is the subtlety of it, the crooked trees, the spreading, flat-topped acacias that would delight a Japanese artist, the silver harmony of grays and browns, the fineness of every unerring line.

The most curious features were the gigantic ant hills, huge, hard-baked cones made by the white ant that is the curse of Africa, eating out every wooden foundation. The cone is built up about a tree trunk, and in some cases the original branches of the trees are still sticking out of the top, but more often every vestige of the original support has disappeared. In Elizabethville we had seen one ant hill, in a garden, used as a storehouse, with a door set in place.

A few antelope could be seen now and some secretary birds, but the country was as barren of game as the veldt, and we all began to wonder if we were really in the interior of Africa when, after a day and a night, the little railroad—built within the last five years—ended abruptly upon the banks of the Lualaba River and our adventuring began.

The Lualaba River is really the Upper Congo, but it achieved a name of its own before Stanley discovered the connection. Bukama, the railway's end, was the highest point on the river to which light steamers can penetrate. Now it was dry season, before the Rains, and we were uncertain about any river transportation, so when our train came to its final stop by a thatched depot at the river brink we rushed out into the tropic dark and discovered that, though a regular steamer was not here, there was a small affair that could take us and would start in the morning.

Our delight was great. It lasted just thirty seconds, until we received the answer to our second question. Our luggage? But there was no luggage.

I never saw anything so sudden and so sickeningly

complete as our despair. We stood there in the raying lantern light and looked tragically at each other and incredulously at the light-hearted Belgian who told us this. He had no conception of our anguish. He went on to say that while we waited for it we could live in the private car—they would leave it on the tracks when the train went back.

But our own eyes, the masculine eyes of the party, had seen the luggage leave on a goods train before we left. That train had arrived. Therefore the luggage must have arrived. . . . But there was no luggage. Grimly Mr. Akeley spoke of trouble. Amiably the agent assured him that he lived on trouble and recounted an anecdote, perhaps to woo us from our private griefs, of a Belgian doctor who had come from Brussels with a magnificent equipment which had been lost and never found, although he had waited five months.

Urgently my husband insisted that the official was mistaken and that the luggage had come but had been overlooked. I began to be ashamed of his insistence. To have overlooked that mountain of luggage suggested such bare-faced negligence on the part of the official who, after all, was not responsible for its nonarrival!

But my husband is an insistent character and he described the luggage and repeatedly detailed its departure in the car, and at last the word "car" had a revivifying effect. There was a car, it developed, but it had not been opened, for the papers for it had not come. So nothing could be done.

But a great deal was done. We had papers enough,

and the official obligingly met us halfway now that the mystery was solved, and the Bukama affair ended in tea and hospitality in the thatched depot and such exuberance of good feeling as only the reprieved at the gallows' foot can know.

We were off next morning after a torrential downpour, the first rain of the season, in an infinitesimal steamer, piled very full with ourselves and the wood fuel and the native crew and the Belgian captain; our luggage in an open scow lashed to one side, and some natives in a similar scow on the other side. We sat forward under a little iron roof and ultimately scrambled up the rail and sat on the top of the roof.

We had made no preparations for food, having been told at Elizabethville that the passenger steamer could provide that, and at the last moment the kindly official in charge of the freight depot came rushing down the bank with the hot and fairly kicking leg of a goat which he had just that minute despatched for us. However, there was another passenger, another river captain journeying as far as his own grass hut, where he and his wife contributed antelope and everything else we would accept, so the goat was a vain oblation, although appreciated by our boys.

All that day we glided between groves of borassus palms, the loveliest trees in Africa, and past banks overgrown with forest, always with purple-peaked mountains against the sky. There were plains of antelopes and there were three elephants tranquilly feeding; there were crocodiles slipping silently from sandbars at our approach; and there was a huge hippo

Borassus Palms on the Lualaba River

[page 44]

Congo Belles Traveling

[page 44]

Courtesy of Major Flint

CROCODILE WITH ARM AND LEG OF NATIVE HE HAD JUST EATEN

[page 48]

GAVIAL, ASIATIC SPECIES OF CROCODILE, CAUGHT IN LUALABA RIVER

[page 48]

swimming open-mouthed toward us. We were a long way yet from gorillas, but we began to feel that we were on the trail, really in Africa, headed up to the heart of it.

Our traveling companions in the native barge were Congo belles returning down river clad in the latest thing in calico and beads. At Elizabethville a native woman with any pretension to fashion wore a regular dress with a yoke of white embroidery and with short, puffed sleeves. Now the girls wore a length of calico knotted under the arms—not any kind of calico, for Zanzibar set the styles and the vogue changed with the advent of fresh traders. A Paisley shawl pattern in maroon was particularly good this year.

Across the smooth, dark shoulders and arms that this classic simplicity left bare, went the darker pattern of cicatrization marks—little cuts made in the skin when a child is young and filled with ashes to prevent healing. On the men the marks were often tribal in significance; on the women they were made apparently only for ornament—little flights of arrow heads crossing the shoulders or relieving the monotony of flat cheeks.

The Congo style of hair-dressing was elaborate; it took a girl an entire day to do her hair, dividing the head area into circles and segments with mathematical precision and twisting the hair strands into black strings that she knotted tightly under her chin. Then, being near enough to civilization to feel the influence of the bandana, she wound another piece of calico about the coiffure and the emerging strands of hair had somewhat the effect of bonnet strings. The native comb was of wood, narrow and deep and decorated with black paint.

That night we made connections with a freight steamer whose captain, frankly amazed at this invasion of Americans, gave us the freedom of the ship and we tucked ourselves away, the girls in the cabin of the absent mechanician, and Mr. Akeley in a room on the accompanying barge. Herbert and I set up our cots on deck. By day we sat forward on the barge in our steamer chairs, gliding along as luxuriously as Cleopatra in a dhahabiyah and through more marvelous scenery than even the Nile can offer. Our three boys cooked for us over an open fire on the steel deck. As all the native passengers camped about the back of the barge had their own little cooking fires, too, we presented the appearance of a floating picnic ground.

We had opened one of our London chop boxes containing an assortment of good things in tins—coffee and tea, milk, butter, sugar, jam, cheese, prunes, deviled ham, tongue, etc., and at our stops at native villages we bought eggs and bananas and chicken—eggs at four cents the dozen, bananas at three and a half cents a bunch, and chickens at seven cents apiece. That was scandalously high for chickens; we rarely paid more than three and a half cents after that.

We had to be careful of Alice on this boat on these slippery, unrailed decks, but we never let her stir alone and found a safe corner where she spent her days at her favorite drawing and coloring. The river was full of crocodiles. Several times we saw them drifting past, like submerged logs, and twice, looking over the side,

I caught a crocodile's goggle eyes looking up at me in sinister expectancy.

The captain of our first little steamer had just had a friend pulled under by the crocs while taking a rash swim. It seems to me the most horrible of deaths. There are several theories about the way the crocodile consumes its victim. Some hunters say that he crunches down his prey at once; others, that he holds it under until it is drowned and then often hides it away on the bank until it is what might be called riper and more flavored to his taste.

Major Hamilton, for many years game warden in British East, is a partisan of the filed-for-reference theory. He vouches for the story of a young man who was pulled down and held under until the crocodile judged that he was drowned. The young man revived in a black, wet cave in the river bank, worn by the waters when high, but just now at the river level. He was uninjured but he did not dare try to crawl out towards the river where the crocodile might be, so he clawed at the earth above him and succeeded in breaking his way through the sod overhead and making his escape.

I haven't the slightest right to an opinion, but after hearing a number of first-hand stories on the subject it doesn't seem to me reasonable that the crocodile is bound by any hard and fast rule. If he is hungry it would be natural to eat at once; if his appetite is satisfied, but he has a chance at something, it would be equally natural for him to pull it down and thriftily hide it away. That he can ravenously bolt his food is shown by the photograph of a specimen shot by Major Flint of Uganda, where the entire arm and leg of a hastily consumed native were excavated from the croc and piled jauntily on his back as a photographic exhibit.

Five days on the Lualaba brought us to Kabalo, a Belgian outpost, where there was an administrator and his wife, the young *chef de gare,* and two representatives of trading stores. Here we camped on the river bank to await the twice-a-week little train to Lake Tanganyika. While here the natives found a crocodile entangled in a huge fishing net and brought it ashore in front of our tents. They had tortured it brutally and it was nearly dead when I saw it, motionless and hardly breathing. It was all the more sickening because it was not a man-eating crocodile at all but a species of gavial, a fish-eater, an Asiatic species not usually found in Africa. The snout is different from that of the true crocodile, thinner, and curving to a kind of knob at the end.

From Kabalo our last reach of railroad brought us in a day's journey to Albertville on Lake Tanganyika. At Albertville I had a letter from a friend sent ahead to greet us. He hoped that we would like Main Street and that the hotel would not be all we feared. Hotel! He had been misled by that "ville" in Albertville. Our hotel was a windy hillside so hard-baked that it was impossible to pound a tent-peg into it, and the first night the men did not get their tent up at all but slept on cots, where morning brought the magnificent sunrise over Lake Tanganyika, that daylight revealed lying at our feet.

Tanganyika is a glorious lake. It is four hundred and fifty miles long and about forty wide; with the exception of Lake Baikal it is the deepest fresh-water lake in the world, for soundings of over two thousand feet have been taken. It is twenty-two hundred and fifty feet above the sea. It is a great stretch of blue water, encircled with a golden rim of sand, backed by rocky headlands topped with a fringe of crooked trees. And behind the headlands, against the horizons, were mountains.

We had no more beautiful hours in Africa than those late afternoons we spent upon the beach watching the women, with their great water jars upon their heads, go back and forth, climbing the winding path to the villages upon the cliffs. It was beauty itself, the perfect loveliness of beauty undespoiled. But it was beauty with a doom on its bright head, for already the white man had come. There were steel tracks reaching in from the west to Albertville, and across the lake there were tracks stretching east to the coast at Dar Es Salaam. In ten years, twenty or fifty, civilization and its gauds would be enthroned. The old Africa would go.

He walked the Tanganyika beach,
A slim young native, unconcerned,
The wash of Tanganyika round his feet,
The roar of Tanganyika in his ears;
Behind the darkening headlands the set sun
Flushed sky and sand with amber and with rose;
The fringe of flat-topped trees upon the top

Was silhouetted black;
He walked serene, wet with the waves
That for uncounted ages had washed out
The footprints of his people on the sands;
He did not know nor feel nor hear
The wash of civilization round his feet,
The roar of civilization in his ears;
He did not know that he was doomed,
He did not know that soon
The cuckoo's silver call would hush before the siren's scream,
Factories would rear their belching necks upon the heights
And poison the sweet air;
The trees that fed his frugal fires would fall for fuel
And turning wheels make gauds and gears for smooth-tongued
men to sell;
Money would clink,
His solitude would be a prison and a market place
Where sweating toil wipes sullen brows
And dull eyes gape at reeling films depicting scenes afar.
He and his race are doomed;
The ancient land is doomed.

The captain "ate" us. That was his word, not ours. It meant that our meals, served on the deck of the lake steamer *Baron Dhanis,* by our boys on our dishes, and the meals of the twenty other passengers—Belgian administrators, fiscal agents, judges, missionaries—that we accumulated and hospitably accommodated upon a boat with staterooms for ten, were all prepared by the captain's mite of a cook in a galley whose confines resembled a Chinese torture box.

His *pièce de resistance* was a pig, which hung fairly complete from the rail the first day out. No Congo boat is accoutered until it has some former live stock

Giant Palm on Tanganyika

[page 50]

Native Boat on Lake Tanganyika

[page 51]

The Tree at Ujiji beneath Which Stanley Met Livingstone

[page 51]

hanging like a thief in irons at the cross arms. That particular pig we remembered vividly. We had the last of him one breakfast, a headcheese effect, served with onions and mayonnaise. Our best meal was the four o'clock chocolate, which was rich and plentiful, and the captain, a genial character from Sweden, used to chuckle at our strange passion for it.

We were five days getting to the north end of the lake, tying up for two of them at Kigoma, on the east shore, in what used to be German East. After the war it was Belgian and then reapportioned to the British. Four miles from Kigoma is the historic tree beneath which Stanley met Livingstone. My great-grandfather used to talk about that meeting, and I had always pictured it in some dark and damp African forest, mosquitoes humming and danger threatening. I wished my grandfather could have seen that place of sunshine and dancing waves!

I heard here for the first time Stanley's exquisitely conventional greeting to the white man he had come so far to find, "Dr. Livingstone, I believe?"

Ujiji, where the meeting took place, is a native settlement of about twenty-five thousand and was formerly a great Arab trading center in slaves and ivory. The first white men to visit Tanganyika—Burton and Speke—arrived here in 1858. The first steamer on the lake was a missionary boat launched here in 1884.

At Usumbura, the north end of the lake, to which the *Baron Dhanis* eventually brought us, we camped five days waiting for the two hundred porters for our march north through the mountains to Lake Kivu. We had come now as far as rail and boat could take us. We had hopes of a boat on Lake Kivu, for a government launch made the round of the lake once or twice a month, and to avoid waiting for it, Monsieur Ryckman, the commissioner at Usumbura, authorized us to use it whenever we arrived, if we would pay for the gasoline—but the gasoline was lacking. However, it was ordered and was somewhere on the road. This lent considerable uncertainty to our traveling, but uncertainty is a condition of safari life. We had left our last telephone far behind at Bukama on the Congo—and that was broken. There was nothing quicker now than the black feet of a runner.

Our wait was enlivened by visits to the native market which was held early every morning in an open space about a mile from our tents. It was the largest and most voluble market I ever saw or heard; the natives streamed in to it daily from every direction; from the forest came the firewood venders, from the cane fields the sellers of sugar cane, from the village shambas the possessors of eggs and grain and meal and chickens, and from the lake the fisher folk with baskets on their heads —everything was carried on the head—heaped with tiny silvery fish, as microscopic as white bait. I never went through the market but I found somebody picking half a thousand of those fish out of the dust after a jostling episode.

Our boys went to this market, without, or against, orders, with a persistency worthy of Mary's little lamb. We had now six boys and a cook. We paid the cook two dollars and eighty cents a month and food money, and out of that the cook paid his helper. Our latest boys we paid only a dollar thirty-five a month and food money. Each boy had a book which we kept for him and in which we entered whatever he drew against his wages. Here at Usumbura the boys were drawing lavishly and one day we found why, and why they loved the market so.

There was a bar—a row of huge earthen jugs holding *pombe,* the native banana beer, and behind the jugs sat the barmaids calling their wares. There were two classes of service; in the more expensive the buyer got his beer in a cup, in the cheaper he had a pull at a straw. Here the technic was for the buyer to draw a deep breath and suck as vehemently as possible while the vender endeavored instantly to choke him off with loud revilings.

We bought interesting baskets here, very simple circular ones with rounded covers that fitted snugly. We never saw this shape farther on. And we bought many of the men's bags and pouches, of tightly knotted fiber, sometimes ornamented with knotted fringe, which the men wore slung over a shoulder and in which they carried pipe and tobacco and food and knife and perhaps firesticks and money. One day I bought a very elaborate new one to the loud lamentations of the man's wife who had just finished knotting it, and now saw herself furnished with more occupation for long days to come. It must have been incredibly tedious and difficult.

Another diversion of our wait was an elephant hunt on which the administrator took us, but in which the elephants failed to coöperate, although we rose at three, bicycled ten miles to a river, and then walked till five o'clock in the effort to meet those elephants half-way. We saw beautiful country, elephant spoor several days old, but nothing alive save the gleam of a leopard or serval cat and the gray and scarlet of the talking parrots.

All the excitement of that day had occurred in the camp after our departure. For some days we had noticed the presence of what might be called Congo flappers about the place, provocative-eyed young things with the latest cut of hair and the latest pattern of Zanzibar calico knotted beneath their smooth-skinned shoulders, the latest sparkle of beads and anklets, and these flappers were lingering about the cook-tent undoubtedly longer than it took to sell eggs or poultry. One was particularly in evidence.

"Jim," said Mr. Bradley, Jim being our abbreviation for Gimway, a white-toothed roustabout of a "boy," "Jim, who is that woman—who is she?"

Jim's English was usually inadequate, but this time he was primed.

"Cook's sister," said he.

"M'm, cook's sister . . . And is she," said Mr. Bradley, "by any chance also the cook's wife?"

Jim could not forbear making the case stronger. "Yes, sare, cook's wife."

"Well!" said Mr. Bradley. He added slowly and impressively, "She must not stay in camp. Tell cook he must send her away. We do not take women with us."

"No, sare," said Jim earnestly. "We take, sare!"

PORTERS ON THE MARCH

[page 58]

ALICE ON SAFARI

[page 58]

NATIVE MARKET, USUMBURA, LAKE TANGANYIKA

[page 53]

NATIVE BARMAIDS AT USUMBURA

[page 53]

Somewhere in the tents Martha and I snickered. But the champion of virtue did not crack a smile.

"No, you don't take," said he, firmly. "We—don't—take—any—women—on—this—expedition!"

Jim looked blank. . . . He must have thought it a white man's world. . . . "Send her away," repeated Herbert, adamantly.

She went away, but when we were off elephant hunting, she came back again, and after her came her husband. He came with a whip, the kiboko of hippopotamus hide, and dragged her out of the cook-tent and began to beat and choke her. The outcries brought Miss Hall from her tent, and she shouted vigorously at the man to stop and go away, but he paid no attention to her nor to the boys. The cook made no effort to interfere on behalf of his lady love and the husband had apparently no animus against the cook but directed all his feeling and blows at the girl.

Priscilla was prompt and intrepid. She got her revolver and marched out with it in her hand. "Tell him to go away!" she said vigorously to Kiani, and the vigor and the revolver had their effect. The man stopped, the girl fled crying to Miss Hall, the boys talked all at once and the man went away. Immediately the girl cheered up and showed every intention of remaining, but finally she too drifted off and the place thereof knew her no more.

But she was only the first of the cook's sisters. It was perfectly extraordinary how our line of march coincided with the spread of his family. Invariably the sisters were about fifteen, with silky-lashed eyes and soft, giggling voices and very, very expensive beads.

Changing the cook did not change the phenomenon. It seemed characteristic of the Congo *impishi* to have sisters in every tribe.

The administrator at Usumbura had sent out to the chiefs for porters, and early on the morning of October 14 we saw the long line of blacks descending upon us and felt that we were on safari at last. We had come now as far as rail and steam could take us, and had nothing now to depend on but our feet.

CHAPTER V

ON SAFARI

A Sudden Elephant Hunt; Martha Gets a Tusker

In the Congo, the Eastern Congo, you sit upon a mountain peak and gaze out to other mountain peaks, like pastels with distance, and you do not wonder how you are going to those other peaks—you know. You walk. And everything that you want, tents, food, clothes, camp equipment, walks with you on the heads of black carriers.

This caravan life is called being on safari. A safari day begins early—four or three, by starshine or moonshine or black night. It takes courage to throw back the blankets—those African highlands give winter-cold nights—and grope for the candle lantern, which has always changed position since you staked it painstakingly at bedtime, and then plunge shiveringly into khaki. As you dress you hurriedly thrust the discards into the green linen bags beside the cot, and your final deed is to roll up your bedding, force that into the bag, cap it with your linen bath tub and enameled wash basin, and add any little odds and ends left out, ere you sally forth to breakfast.

The table is outside in the blackness of tropic morning, candles glimmering wanly upon it. Before you are through with the prunes or the pineapple or the oranges you are in a gray gauze world, from which veil after veil of gauze is being swiftly withdrawn, and by the last bite of bacon or burnt toast and marmalade, a scarlet fire is streaming over the mountain tops and a rush of light is mocking the feeble candles.

Behind you the tents have been taken down and rolled up in bundles for carrying, the bags and boxes brought out, and everything made ready for the march. The invariable last thing is the breakfast dishes, which the boys do unhurriedly, conversationally, and bring, in trickles of two or three, while you stand guard over the empty boxes and keep off the porters, who yearn to acquire the boxes in their empty state and make off with them.

Our porters at Usumbura came in four lots, from different villages, with a headman from each village. The porter's status seems to be that of a voluble chattel. I don't believe any one ever volunteers for the position. In the Congo, porters are paid half a franc a day, which was three and a half cents for us, and ten centimes a day for *posho,* or food-money. The day's march is about four hours, though often more, and I have known it to be nine, and each man carries about forty to sixty pounds on his head. Our chop boxes were all made up in weights for carrying, and we tried to have all the loads of equal weight, but of course there were discrepancies.

There was a strong interest taken by those porters in the loads they got that first morning of our march, for each man carried the same one for the rest of the trip. With an unerring instinct for light boxes, they surrounded the fly tent where we were breakfasting in the gray dawn and the meal was made hectic by having to whirl about and frustrate the long, black, snakelike arms reaching under the canvas to purloin some unready box. We finally got under way about seven-thirty the first morning, with a triumphant blaring of the curved horns the leaders loved to blow. The music resembles a Scotch bagpipe crossed with a mule, vocal, and a really musical safari *en route* sounds like a calliope gone mad.

We Bradleys went ahead, to lead, or rather discover, the way, the five-year-old Alice in her basket carried by two natives. After us came the long line of blacks, two hundred of them, a picturesque frieze against the sky line, in girding goatskins and beads, on each high-held head its load. Mr. Akeley followed at the end, to see that no loads were left, and to have the opportunity of taking motion pictures without halting the safari. We had bought bicycles at Elizabethville, for no mules were to be had, owing to the difficulty of bringing animals in through the deadly tsetse fly-belt, and though we knew nothing of the route, except that it mounted over two thousand feet in the eight days, every bit that we could ride would be a help. For Alice, Mr. Akeley had devised and practically built a seat low on the front wheel of one bicycle, and that was ridden by her boy Mablanga who was strong, reliable, and utterly devoted. She loved this way of traveling more than any motor, at home, and much better than the basket, carried by two natives, to which she was generally consigned.

When riding was impossible, the porters carried the wheels on their heads. It had been said that one can ride all over Africa on a wheel, for the native paths, padded smooth by generations of bare feet, wind from village to village. In the administered territories the chiefs are required to have a certain amount of work done on the paths, cutting the side grass and so forth, so the ways are improved, and even a few miles of riding on a long march is tremendously worth while.

I had no idea that I was a good rider—indeed, I had no idea that I could remember how to ride a bicycle at all—until I had the choice between riding down a mountain and walking. Then I took goat trails and dried courses of brooks, rock slides and thickets with a dash and abandon that covered me with bruises, but certainly covered the ground.

Our first day's march from Tanganyika was a long one, twenty-three kilometers, and the road that day was good, so we bicyclists reached the rest-house, Koto-Koto, at noon and sought shelter from the sun under the thatched roof, where it was always cool, and proceeded to wait for the safari. We had brought water bottles and some lunch in a bag, and we bought bananas from the chief, who promptly presented himself with the inevitable present of a chicken.

These presents were a trial from the moment they arrived, heads hanging, legs tied, to the time of their assassination and ultimate appearance upon the table. To become a present, a rooster must have lived long past all reasonable expectation of a Methuselah and apparently subsisted all those years upon what sustenance he

Personal Attendants on the March—Camera, Gun and Wheel
[page 60]

Alice and Mablanga

[page 60]

Alice and Her Tent

[page 61]

Porters Waiting for Rations

[page 61]

could pick up within the radius of a two-inch tether. Nothing stringier or more sinewy exists.

In the Congo you shake hands with a chief twice, once pressing the hand and once the thumb; so we each shook hands solemnly, even the five-year-old, whom he greeted as the Sultana with special consideration. Alice's presence with us was a perfectly simple thing to the natives; evidently Alice and I were favorites and if I left her at home the other wives would naturally poison her! The mother of Musinga, the great king of Ruanda, is credited with having poisoned six of his little half brothers to clear the throne for her son.

Our tent boys understood the relation of the party and the white man's ways, but to the natives I was always the "Wife-with-the-Child." One chief asked me how many wives my husband had at home—he himself had sixty-five around his different villages—and I really felt I should be imperiling my husband's position in his eyes if I attempted any explanation. I have never yet revealed just what I did say, but that chief was impressed.

That night and several others we slept in the grass rest-houses built for traveling Belgian officials, for these looked fresh and there seemed no danger from the dreaded kimpootoo or spirillum tick. We always had our nets over us in those houses, although we rarely saw a mosquito. The freedom from flies and mosquitoes was one of the joys of Africa, contrary to every one's belief about it. Sleeping in the houses did away with putting up tents, and when the porters came marching in, horns blowing, although the sweat was running down their black skins, we had only to see to sorting the stuff.

The moment a porter arrives he sheds his load and rushes to find a sleeping place in one of the huts. Your duty is to stand over the loads and see that they are left in order and the bags sent where they belong. A headman does this for you, but there were no safari headmen in the Congo, and we all pitched in and did a little of everything. About the time camp is settled and the bags and boxes distributed, the porters have brought in firewood and water, the cook has his fire going between the three stones, and you unlock the food boxes and hand out the precious rice and butter. Curried chicken was the usual mainstay of our meals, with rice as only the East can cook rice; if in camp, we invariably began with soup, but on the march our first course was *hors d'œuvres* of sardine and coldslaw. We could carry cabbage for days, and we had potatoes, both white and sweet. When no fresh green vegetables were available we had asparagus in tins or fried bananas. Our dessert was any fruit available, pineapple or papaya, with crackers and cheese and jam. The Congo variant of pancake with honey or jam was another resource. By the time that dinner is under way the chief is generally ready with *posho* for the men, usually plantains or coarse meal tied up in banana leaves, and the porters line up opposite and voice their objections.

Life among the porters is always a commotion, but *posho* time sounds like a riot. We grew very casual about noises. An outcry that would have blanched us in America did not turn our heads. The conversational qualities of porters are unequaled even by Venetian gondoliers, but actual fighting is rare. We had only one experience on the way. After the porters get their food they go back to the grass huts and wisps of smoke curl up from many fires and the chatter, chatter goes on, unending, in that strange tongue, which, being meaningless to you, becomes a part of the day's background. And you eat your own meal out of doors, in the lovely hour of waning light, and sit and watch the sun sink like a flaming disk behind the mountains, while the crickets and night noises begin, and a swift darkness pours over the land and the stars blaze overhead. And then you go to bed remembering that at four you will hear a call.

From Tanganyika to Kivu we followed the Rusisi River, which drains out of the southern end of Lake Kivu, and we crossed that river sometimes half a dozen times a day. The way of our crossing was simple. There were no canoes, no bridges, and the current was strong. We went over on the backs of the nearest natives—our own boys, if no one else was available, but generally we impressed a near-by loiterer. The first time I mounted I felt the qualms of the first man that ever ate an oyster, but in time I learned to survey my mount critically for strength and endurance and then spring lightly upon his shoulders and steady myself by his topknot.

The country through which we traveled was beautiful, always among mountains, with wide outlooks over forested valleys and plains. There were tremendous numbers of the dark, striking euphorbia trees and of the beautiful scarlet-flowered Kaffir-boom, which we had seen all along the way from the Cape. The branches were invariably bare, a silvery gray; and a network of these trees, gray branched. with brilliant flowers flaming against a sky of burning blue, is one of my vivid memories.

On the second day's march our path was crossed for the first time by safari ants. These are the traveling scavengers of Africa, crossing the continent in endless streams, consuming every dead and living thing that falls into their clutches. When the column pours through a native village, the villagers in the line of march decamp at top speed, and the invaders render them a valuable service by cleaning up the place, from fleas to refuse. They can kill a rat and pick it to the bones, it is said, in two minutes.

The column we first saw was like a black ribbon winding across the path, a ribbon about four inches in width. It was a living, seething stream of ants. Along the sides were the soldiers, big fellows with pincerlike jaws, holding the lines, preventing any outpourings and attacking any enemy, while down between the lines, in a tunnel-like groove, the unending army poured along in such dense crowdings that some were scrambling on the backs of the others. Straws that we dropped across the line were instantly seized and maneuvered into position and carried on out of the way. When we stood a foot or two away the soldier ants did not bother us, but as we moved nearer they streamed out with menacing little jaws. When we poked at them with a stick some went for the stick but others went straight for our boots and up our puttees.

A bite is a vivid sensation. I have seen a hardened porter, one who stood about on charred wood fires until his feet smoked without appearing to experience any discomfort, leap like a deer when a safari ant got him in its bulldog grip. The numbers seemed endless. We watched that column for two hours that first day not knowing that three times more that day we were to cross them, and that scarcely a day thereafter would we be without a sight of them, and all those hours the stream of mites was like a racing mill stream of black waters surging down that groove. Thousands—millions—billions. I could understand why sleeping men, surprised, had been swiftly bitten to death.

Once, at Lulenga, a column passed near our camp, continuing for seven days, and not for an hour at day or night was there any cessation of that seething stream of them. But, seen in time, there was no danger. One could step over the column and be safe.

It was the third day out that Martha Miller killed her elephant. We were in camp on a high bluff when the elephants were seen feeding in a plain about two miles away. It was four-thirty, and the light would be gone only too soon, so we set off, with guides, guns, and cameras, at a grilling pace. We reached a hill above the plains and crept cautiously forward through the bush till on the brow we had a splendid view of the elephants. There were six of them about two hundred yards below us, five full grown and one smaller, all feeding utterly unaware of us, their trunks swinging in the bundles of grass, their tusks gleaming against the dark skins. There is something prehistoric looking about a wild elephant that makes him seem unreal. He is a left-over from another age. His very size is archaic.

They were a magnificent sight, the six of them, and we were able to watch them under unusual conditions. It was a good enough light for a picture, Mr. Akeley thought, and we crouched behind bushes and watched while he maneuvered forward and took some film. The light was going fast and the elephants working away from us across the plain. If we were to have any shooting it would have to be in a hurry, so we crept forward after Mr. Akeley, until we were about a hundred yards from the dark shapes that were growing darker every instant. Mr. Akeley picked one for me to shoot at and one for Miss Miller. The men were to follow our shots with their heavy guns.

When he said "Fire," the bull I was trying for was swinging about in the brush, and my only chance was the head shot. I tried to place it low, guided by the glimmer of the tusks, and I thought I had succeeded, for he staggered, and I thought he was down, but he recovered and made off, although my husband and Mr. Akeley opened instant fire and Mr. Akeley was sure he had placed five shots in the animal.

Through the cannonading I heard Martha saying, "Mine's down," and so he was. She had placed a splendid side shot through the ear. She started to fire at the black dots of escaping elephants when hers got up again, and was then brought down for good. We cut about in a circuit and made our way to him. He seemed enormous as he lay there. There was something sad about the helplessness of his great bulk. The only reconciling thought is that the elephant is doomed anyway, and this was a quick death.

It was so dark that we turned back immediately to camp. We heard the wounded elephant trumpeting and believed we should find him dead in the morning. I hoped so, for his sake and my own, for I hated to think of a wounded elephant coming to a slow death, and, now that I had wounded him, I wanted those tusks of his. But the night rains washed out every vestige of track, and the natives sent next morning to reconnoiter could not be stimulated by any offer of reward. They circled about and ended near the other elephant, where they wanted meat. But, however much we regretted the escape of Adoniram, as I christened my lost elephant, we were elated over Martha's good luck, and had a dinner that night in her honor.

The men spent the next day cutting out some skin and the ivory. It was a single tusker, weighing about fifty-eight pounds. The other tusk had been broken off from an old bullet wound, and there were sores from wounds on his body. The men had a hard time getting work from the natives, who were interested only in getting meat, and whenever the white men were obliviously occupied the men would sneak up and hack off choice morsels, to be conveyed by the grapevine route to the outskirts. There was one leprous-looking old chief who had a huge wash basin and a total disinclination to work, and one of the brightest moments of the day was a sudden vision of that chief sailing through the air, followed by his basin, as my husband turned and caught him making a felonious assault upon the elephant.

No work, no meat, was the order of the day, until the elephant was turned over to the chiefs and headmen for distribution. There was feasting and dancing all that night and the dark was lit with little fires over which the meat hung drying while in the smoky gleam the dark shoulders jostled and bare feet stamped in rhythmic frenzy.

The next day we went on, abandoning the wounded elephant, something which we would not have done under usual conditions; but until the gorillas were secured we could not afford to waste time on anything else.

That day the men were wild and hilarious after the dancing of the night before and refused to go further than an old rest-house, which we reached at ten-thirty in the morning. The rains, which were happening vehemently and regularly every afternoon or night now, had turned the red clay paths to indescribable glue, but another night of rain wasn't going to make them any easier. The offer of double pay had no effect, so we put up our tents, as the house was old enough to be unsafe because of the danger from infection. The porters went gayly off for an antelope hunt.

Soon we heard them coming back in angry excitement, dragging a man along with blows and yells. One of the porters had a cut in his forehead, made by this man's arrow, and he said the man had attacked him purposely. The man said it was an accident. That was as much as we could make out with our inadequate Swahili and our boy's fragmentary misinterpreting. The prisoner was not one of our porters but one of half a dozen carriers coming through with gasoline for the

Kivu boat—the gasoline we were delighted to know was arriving.

The only thing to do under the circumstances was to take the man along under guard to the white administrator at Kivu, four days away, and we told the headmen to guard him and keep the men from hurting him, and a little later Herbert and I went down to the huts to see what was happening. The porters were swarming like angry bees in front of the prisoner's hut, and two hundred angry porters sound like a Russian village in its daily revolution.

The hut seemed to be full of them, so Herbert and I crawled in to investigate and found the prisoner, moaning and crying, tied with cruel tightness, while about him bent several blacks in the act of desisting from whatever they had been doing before our arrival. We laid the law down again, had them give him food and drink and tie him decently, and ordered the headmen to stand guard and be responsible for keeping the rest out, promising to come down in the morning and see how they had followed instructions.

That sounded all right, but we hadn't the slightest force to back us up. In British East a hunting safari has a couple of armed askaris to maintain discipline. We had our headmen, wild defiant creatures, and a tiny black soldier from the garrison at Usumbura, who was a child in the hands of those headmen. There was a furious storm that night, the thunder muttered and rolled and crashed and the two hundred porters' voices rose and fell with mutterings and roarings like the thunder till I went to sleep at last.

I woke suddenly; out in the blackness among those porters, some one was screaming horribly. It was over in a moment; the shrill high voice was drowned in a babel of frenzied sound . . . It did not come again.

I didn't know what to do; to go and rouse Herbert would be to have him go down alone among those two hundred crazy blacks. Mr. Akeley had gone to bed before dinner with fever. While they wouldn't have the courage to touch a white man in the daytime, there might be danger, at night, from a spear in the back. Many of them had the filed teeth of the degraded cannibal tribes. And, whatever was done, was done. And very likely nothing was done—the porters always sounded like a riot.

I tried to dismiss it and after a time I went to sleep again. In the morning the man was gone. There were three stories given us to explain this. The first was that he had escaped in the night, without any one's discovering it. That was a remarkable achievement, since the man was tied and the headmen were in the tent choking the entrance, and the entire safari was awake all night. The second story was that he had just that moment got away, and several of the porters with spears were staging a little retrieving party and running about the woodland looking behind rocks and bushes for as much as ten minutes. The third story, told furtively by a frightened boy, was that they had killed the man in the night and eaten him.

We never knew.

CHAPTER VI

THE SUMMIT OF AFRICA

LAKE KIVU; THE GORILLAS AT KATANA; IN CAMP AT KISSENYI

WE stood on a hill top overlooking Lake Kivu, feeling we were looking at last upon the promised land. For months we had dreamed of seeing this lake, reported the most beautiful in Africa. It was the last to be discovered of the large central African lakes. Rumors of its existence had been brought to Livingstone at Ujiji by Arab slavers, but the first accurate knowledge of it was obtained in 1894 by Count Grotzen, who visited its northern end. We were the first Americans to reach its shores.

All that day, the eighth of the safari, we climbed endlessly, higher and higher over the last great barrier of mountains. We had crossed the pastoral basins on the highlands, an African Switzerland where we wound among grassy slopes dotted with hundreds of native cattle, picturesque black and white creatures, with wide branching horns; past lush marshes, where the golden-crested crane were preening; past shambas of bananas rustling their suave fronds—an upland paradise for the dwellers in the little grass huts—then on, up and down and up again in switchback progress among the hills.

They are everlasting, those hills which surround Kivu,

and they are extraordinary. They are grass-grown, innocent of forest or deforesting, curving crests of velvet softness between the shadowy mountains and the gleaming lake.

Kivu has a beauty like nothing I have known. And I saw it first in a moment of sheer magic--a black storm cloud scudding across the lake with a purple drive of rain into waters that held every answering gleam of storm and sky—purple and jade and peacock blue against a silver mist of cloud shot with rainbows; and beyond the storm, sunshine—sunshine ineffably tender, ineffably radiant, shining upon those fairy hills and enchanted mountains.

The lake is the very summit of Africa, almost five thousand feet above the sea. To the north the country drains into Lake Edward and so to the Nile, to the east into Victoria Nyanza, to the west into the Congo basin, and the waters of the lake itself run south through the Rusisi River down to Tanganyika.

The lake is deep, over two thousand feet in places, and there is not a crocodile in it. We went down to the southeastern end, to Changugu, and sent a runner for the little launch, which was across at Bukavu.

Changugu is simply one of the tiny outposts that keep the white man's way open--a garrison of about twenty native soldiers and a Belgian officer. But Changugu was more than that. It was a home. It was Lieutenant Keyser and Madame Keyser and their little son, Jean-Jean. They received this astonishing invasion with such cordial hospitality that our cook took a complete vacation and the children played so happily—despite the difficulties of language—that when the launch came they parted with frank tears.

The launch had space for our luggage in the hold, our boys in the bow, and our steamer chairs under cover at the stern. We were three days upon it, for the captain traveled only in the morning on account of the storm which came each afternoon. It was the last of October now, the height of the fall rains.

The first day we tied up at Bukavu, on the southwestern shore, and camped upon the banks. The rain did not come until late, and all afternoon the native women danced before our tents. They danced in a half circle, advancing, receding, swaying, clapping, all to a monotonous singsong of weird chants, while the few men who danced did a muscular solo performance before them.

It was jazz in its own home town. It was not the shimmy, but it was a recognizable toddle, and there were some other things that even our younger set haven't discovered yet. There was one little offering entitled "The Dance of the Young Girls," which I piously hope was misnamed, for half the participants joggled through it with "totos" (babies) slung in goat skins on their backs. They were not enticing looking young girls and they were clumsy appearing in the goat skins they wore bunched about their waists out of deference to European ideas. They were the Wania-Bongo. But there was one remarkable woman, a visitor, the wife of the cannibal chief, Kabaka, one of the Waregga.

She was of royal blood herself, the wife who gave the chief his heir, they said, and she was certainly the most royal looking creature I saw in Africa, with the lines of a Greek statue and the perfect aplomb of one, too. She discarded her goat skins and appeared in her royal regalia of copper armlets and leg-rings, with beaded necklaces and a trifle of girdle. Her skin was not black but a gorgeous coppery brown, and she had the carriage which is supposed to belong to a duchess but which is really the perquisite of carrying a huge water jar in calm balance on your head all your life.

And she had filed teeth, one of the signs of cannibalistic habits. It is denied that cannibalism exists now to any extent among the natives, but I heard that day of a village back of the mountains which had made a raid upon another village, and, in the moment of excited triumph, consumed its natives. In the light of that information, and with the memory of the missing porter, I looked upon all filed teeth with a certain gruesome interest. At the least they were a symbol of the good old days!

A Congo administrator told me that fifteen years ago he used to see slaves destined for eating led from hut to hut for prospective purchasers to chalk out upon the living flesh the particular cut they desired. Of course that was only in certain tribes, and it was fifteen years ago, and where the white man has power it is finished. But back of the mountains is still black man's Africa, and little can really be known of what goes on there.

It was the business of the young *agent territorial* at Bukavu, Monsieur Massart, to travel from tribe to tribe, sometimes for months at a time, visiting the chiefs, persuading the refractory, strengthening the bonds of peace, personifying with all his power that vague but persuasive thing—the white man's administration.

He invited us to accompany him and we should have dearly loved to do so, but there were the gorillas to get and Mr. Akeley's time was limited. So on we went with the boat next morning and stopped that night at Katana, a mission of the White Fathers, high on a hill on the western shore. Here we got the story of the gorilla invasion which had occurred just two weeks before.

For the first time in the memory of the mission the gorillas had come down from the forests on the mountains and had entered the plantations of the natives, where the natives found them in the morning busily feeding.

Now, that tribe has a superstition against killing a "man-ape," believing that the man who does so will have childless daughters and that his son's wife will lose her son, so the men were chary about attacking the invaders. They tried first to drive them off with drums and shouts, and then, at the chief's orders, went in and tried to scare them off with sticks.

The female did make off, apparently, at their approach, but the male turned on them, tore one man to pieces, and wounded four others. The rest of the men rushed in with their spears, piercing the gorilla through and through. The skin was so badly torn that it had not been preserved; Mr. Akeley was sorry not to get the skull.

He had some thought of stopping here and starting off towards those mountains for his gorillas, but the rumors about them were too conflicting and the distances too vague. It seemed wiser to keep on to the original objective, where the natives knew the gorilla range and had no superstition about hunting them.

We heard another story about gorillas from Monsieur Dargent, a Belgian administrator on the little boat with us, going out to his post at Sake at the northwest end of the lake. He said that fourteen days' journey in the forest back of Sake there were gorillas which had so terrorized the natives that several villages had been abandoned. He had not seen the gorillas himself, but he had heard many stories from the natives.

That sounded like the good old demon gorilla of Du Chaillu's reports which attacks man on sight and carries off woman. But Mr. Akeley was unshaken in his conviction of the animal's natural inoffensiveness towards man. I could see that a gorilla was going to have to treat him rough to make any impression!

And, of course, these were only native rumors, without any of the circumstances being given. It was perfectly possible that a food shortage had led the gorillas to prey upon the village shambas, and where men and gorillas were brought into collision over food there was no question but that the gorilla was a terrible antagonist. The great length of his huge arms gave him a tremendous reach and he was credited with being able to scoop out a man like a soft-shell crab.

The question was, did he do this only in self-defense, or did he do it wantonly, and did he hunt man and attack him on sight? These were the questions on which we hoped for experience to throw a deciding light. A little time more now and we should put our luck to the test.

It was at Kantana that I began having the fever. Ever since entering the Congo it had been a question whether to take quinine regularly every day or not. Almost all Europeans in the interior took five grains a day on the theory that if you had enough quinine in your blood a stray bite wouldn't infect you, but there were others who did not take any at all until they felt ill. I rather liked that latter theory. I felt so supremely well that it seemed useless to take anything until I felt the need of it. I did give two and a half grains a day to Alice to make her doubly sure and Mr. Bradley and the two girls took their five grains each day, but I did not.

All I know is that I didn't take quinine and I did have the fever. And the same thing later happened to Mr. Akeley, who didn't take quinine regularly, while the others who did had none. That may not mean anything, but that is the way it happened.

I must have been bitten at Usumbura or on our march up from Tanganyika, for Kivu is delightfully healthy and free from all mosquitoes and flies. Now a small boat is no place on which to have the fever. We three Bradleys had our cots on the deck of the launch that night, while the others camped at the mission, and next day I huddled in my steamer chair, gazing heavily at the fairylike shores of Kivu, and thinking that if no one knew the difference between sunburn and a temperature of a hundred and four—it felt a hundred and forty!—I wouldn't enlighten them until I had to. The fever was accompanied by a headache which grew positively thunderous, and when the real thunder broke over us—it came early that afternoon and caught us out on the lake—it seemed a mild echo of what was going on within.

The storm that day was a tremendous gust of wind and rain, and the ship's engine broke, so we rolled about helplessly until it consented to go on again. We were a drenched-to-the-skin lot, in spite of raincoats, when we finally landed at Kissenyi, at the northeast end of the lake.

There, against the horizon, were the far, cloud-wrapped peaks of our objective—Mikeno, Karisimbi, Visoke. We were to camp at Kissenyi until we could collect porters, provisions, and information to go on. Kissenyi has the air of being quite a place. It was built by the Germans, for we were now in the kingdom of Ruanda, in what was formerly the very western frontier of German East Africa. The boundary between German East and the Belgian Congo used to run southwest to the lake, and a little north of Kissenyi; the Belgians maintained a post at Goma. Now the Belgian occupational post is Kissenyi. Along the shore runs an avenue, straight as a marine parade, shaded with eucalyptus and edged tidily with lava rock and shrubs. To the right were the clustering roofs of the native village and the soldiers' compound. There were two white families at Kissenyi, the *chef de poste* and his wife and the doctor and his wife. The doctor was a bacteriologist in charge of a laboratory investigating the rinderpest and other diseases. We had noticed their house coming in—a pretty one of brick in a lovely rose garden.

We pitched our tents in the driest place possible, with the lake at our feet and the volcanic peaks all about the horizon. Our nearest neighbor, Chaninagongo, had a heavy cloud of steam overhanging its crater, colored with ruddy lights; but the crimson glow which burned in the sky at night came from Nyamlagira, many miles away.

All I saw of Kissenyi for some days I saw the first night, for I went to bed and had it out with the fever—and Monsieur Van Saceghem, the bacteriologist, came twice a day, and he and his charming wife did everything in the world for us, sending us eggs and butter and milk each day, and filling my tent with roses.

It was great good fortune to have him there, and it was a distinct relief when his blood tests told me I had only the regular African fever, from an infected mosquito, and not the kimpootoo or spirillum fever, which comes from a tick bite, and is a recurrent blaze, often deadly, with danger of paralysis. The kimpootoo is found chiefly on the caravan routes, where the ticks are brought in on the carriers' feet, and for that reason old camping grounds are not desirable.

But even malaria isn't so cheery while it lasts, and I felt that it was no moment for the news which reached us of a Belgian couple who had gone through the Congo just ahead of us, Dr. Deriddar and his wife. They had gone out to hunt on the Ruindi plains, where we were going after we had finished with the gorillas, and there Monsieur Deriddar had died of the fever and his wife was lying desperately ill, mauled by a lion.

I made up my mind that if I lived through the fever

I would avoid lions—at least avoid going into grass after them, as Madame Deriddar had done. She had been finishing the hunt alone, after her husband had gone back to the camp, feeling the beginning of fever; she had come on several lions in a gully, had wounded three or four, who apparently made off, and in following them up she had been surprised by a wounded one, who sprang on her out of the grass. He had mauled her and then drawn back, and she had had the pluck to get out her revolver and finish him.

I added to my good intentions the firm resolve to carry a revolver in lion country—and oh, how poignantly I remembered that resolve one night upon the plains! But lion country or fever country or any country, I was going to take my five grains of quinine a day. I was unutterably grateful that I had given it to Alice, who was blooming like a rose, playing on the sand on the beach—the only beach on Kivu—and keeping Miss Hall extremely busy with a violent little pastime called spearing gorillas, which she played with native spears.

We were very social, those days of waiting at Kissenyi. While our friends at home were doubtless picturing us as sweltering in some lonely, pestilential jungle, we were enjoying a climate of June's loveliest, untroubled by flies or mosquitoes, and going to tea at the *chef de poste's* and to dinner at the doctor's in true European style!

Our own table was greatly enriched at Kissenyi by vegetables and fruit from the Mission of the White Fathers at Nyunde, three hours' journey back in the mountains. For ten francs our boys brought back huge baskets full of potatoes, leeks, onions, cabbage, celery, and lettuce, and we had oranges, strawberries, citrons, and limes. Pineapples we had not seen since Lake Tanganyika. Papaya—a sweet, native tree fruit, in substance like a melon—and bananas we were usually able to obtain from the natives, and there were always chickens and eggs to be had and often milk—although we preferred our own tinned Ideal milk.

We had another neighbor who had been more than kind during my illness, a young Englishwoman, Mrs. T. Alexander Barns, the wife of an English hunter and collector. She was waiting at Kissenyi for her husband to come down from the mountains, where he had gone for gorillas for a group for the British Museum.

Odd that two men, one for an American institution, one for an English, should have happened on the same spot in central Africa, for the same purpose at the same time! Mrs. Barns and I hoped, humorously, but fervently, that there would be enough gorillas to go round!

And the accounts which came down from the mountain of the cold and nettles, and the crawling on hands and knees made us realize that this was not exactly child's play on which we were engaged. I don't wonder that every one in the country thought the determination of women to see wild gorillas was distinct lunacy. But crawl or climb, Martha Miller and I were going on with it. We had not come the long trail to be turned back now, and if there was no chance on the mountains, Herbert promised he would take the fourteen-day trek to Walikali to the gorillas Monsieur Dargent had told us about.

But our greatest sociability in camp came from our native visitors. Rwakadiga, the chief of Kissenyi, was an inveterate caller. Every morning, just the busiest moment of it, I would see the long procession of spears headed for the tents, Rwakadiga and his attendants in the lead. Then I would snatch at my phrase book of Swahili, the almost universal language of the Arab traders from Zanzibar, and scan it hurriedly in hope of injecting something in the nature of conversation into the ensuing call.

My book was a hopeless affair. It abounded in such sterile offerings as "The foreigner has not received his canoes. . . . The wizard has cheated the ten simpletons of their spoons. . . . My blister is not your boil." But I had added more useful gleanings to it, so that I was able to conduct a strictly centralized series of remarks concerning my desire to buy the skins of animals and small, clean, and beautiful baskets.

The chief would head for the first tent, shared by Alice and me, and we would come out and shake hands ceremoniously with him.

"*Jambo, Mamma*" (Greeting, mother), was his speech.

"*Jambo, mfalme*" (Greeting, chief), was mine.

I was Mamma to all the natives, except the tent boys, who said miss indiscriminately. Father is different in the different languages, but Mamma is the same, Africa over.

Alice had always to trot out and shake hands with the chief, then she would trot back to her dolls or drawing. She was neither afraid of the natives nor particularly interested in them. She did not see the strangeness which we did, or think them at all unusual. They were simply part of the amazing world which she was daily discovering and a natural part of it to her.

They were not at all frightening or repellent, but she did get tired of the way they hung about her tent, sometimes trying to touch her yellow curls. Some of them thought the curls were ornaments and wanted to see if they would come off; others simply wanted to feel that strange, fair hair. But they were never intrusive or impertinent, and brought her gifts of baskets or eggs. None of them had seen a white child before.

One day I saw a boy drop on his knees. He muttered over and over a lot of unintelligible things in which I caught the words Jesu Christ. I called the invaluable Mablanga.

"What this boy say?"

Mablanga took off his cap, his invariable respectful prelude to speech, and smiled in an abashed yet amused way. Then, as usual, when stuck for an explanation, he resorted to negatives.

"No understand, miss."

"I think he says 'Jesu Christ.'"

"Yes, miss. One time he go mission. He see little thing—made like baby. He say that Jesu Christ."

The boy had seen the colored image at a distant mission and to him Alice was the living image. So presently he went away and came back with a little basket for a present. We thought it very touching and thanked him fervently. But he was not satisfied. He made several emphatic remarks.

"What he say, Mablanga?"

"He say now time Jesu Christ make present," responded Mablanga.

So Jesu Christ gave him fourteen cents and he went away reconciled.

Rwakadiga had some distinctly advanced ideas. He wanted a white wife; he had heard there were many in the lands far away and he wanted the white people to write and get him one. He had sixty black ones, but he was willing to part with them—or most of them—for he understood the Europeans—all whites are Europeans—were conservative, and his men would build her a house in true European fashion.

He had suggested this both to Mrs. Barns and Madame Van Saceghem, and now he endeavored to convey it to me, but my phrase book was unequal to anything *intime.* I had heard what he wanted and I understood his words about writing a letter, but my boy was too scandalized to translate his request any further.

There was no hint of race problem in Africa—not so far as white women were concerned. Not one of us ever had the feeling of personal timidity.

We made another acquaintance in Kissenyi which we would have been glad to keep at a greater distance—the jigger or burrowing flea. This is not the jigger of the American woods. It is a microscopic insect, like a grain of black pepper, that without any social formalities of introduction dives into your foot, or more rarely your hand, preferably about the nail, and digs himself in for the winter. He—or I should say, she—lays a long chain of eggs. There was no pain attached to the entry, but at the egg stage the dugout has become a purple, painful blister, and this is the cue for an immediate extraction. There was an Englishman once who wished to take his jigger back to his physician intact in his toe for scientific inspection of the phenomenon; that Englishman died. This burrowing flea is not native African but was landed on the West Coast in the dirt ballast of an old Portuguese trading ship, and carried across the continent on the bare feet of the carriers.

Our first jigger was an event. Repetition staled the interest and evoked aversion. The mud-thatched magazine at Kissenyi where our goods were stored was loaded with the pests and after Mr. Akeley had unknowingly entered the warehouse and then returned to camp we had an invasion of them. Old market squares and native villages are their usual abiding places.

There were no crocodiles in Kivu, and one could enter the waters without fear. At Kissenyi there was the only strip of beach on the entire lake—the name Kissenyi itself means "sand"—and after four-thirty, when we could take off our helmets, Martha and Alice and Priscilla and I had some delightful swims.

Early explorers had stated there were no hippos in Kivu, but at Bukavu Monsieur Massart told me that he had seen one between Bukavu and Katana, and that he was going out some day to kill it—and as my French was unequal to the work of persuasion I expect the last rare hippo has ere now disappeared.

Of the climate of Kivu, every Belgian said, "It is Europe." The temperature is exceptionally equable, one day there rarely varying three degrees from another the year around. The morning is spring, the noon July, and the afternoon spring again—the night October. Every night in Africa we slept under blankets, except for a few hot ones on the *Baron Dhanis* on Lake Tanganyika, in staterooms close against the high bank to which the boat was tied. The only variation in the Kivu climate is the alternation of wet and dry.

The rains, which had begun the first day upon the Congo, September 20, were in full downpour now. The mornings were sunny and cloudless, but between noon and two o'clock a black thunderstorm would come scudding across the sky and the heavens would open with peals and crashings and all the water in the world, apparently, would come down. Sometimes the rain would last an hour or two; sometimes twenty minutes. We found the dampness pervasive, and tried to keep clothes well aired in the sunshine. When I was up after the fever I spread out the contents of my green bag, which had been standing on my tent floor untouched for over a week, and found that the articles at the bottom of it were green with mold—baskets and coin purses and a pair of my slippers. But that was something that a resident with a fire or a dry floor could avoid. These rains, called the light rains, lasted till the middle of December, then, after a dry month, were two months of heavier rain, and this was the usual yearly program.

For mildness and healthfulness of climate, for fertility of soil and for enchanting beauty, Kivu is unexcelled among the Belgian outposts. It would be an ideal place for a colony of artistic and literary folk—for any self-sufficient group who wanted peace and beauty and leisure and the comforts of life without spending a fortune for them. The fortune would be spent on the way in. After that, life at the present cost of boys and food could be maintained on unreasonably little.

When we were in Kissenyi there were only two white families there—Monsieur and Madame Wera and Doctor and Madame Van Saceghem. It is those lonely couples, those exiles on the far outposts, that pay the price of colonization. Every home that I went into was a little bit of Belgium. Against the grass wall were innumerable photographs of the family—all the relatives, big and little, and the pet dogs and kittens. And almost always there would be a series of pictures of some little girl or boy—the child who was left at home. That was the tragedy of the Congo and of all far colonies. It had been a joy at Changugu to see the little seven-year-old Jean-Jean there with his parents. At Bukavu the administrator and his wife, Monsieur and Madame Mignolet, had two little babies, a tiny one in arms and an older brother in pink rompers. . . . A little later, as we left the Congo, we heard that Madame Mignolet had died very suddenly and the stricken husband was trying to get leave to take his babies home to Belgium. The Dargents, who had been on the Kivu launch with us, had been separated for many years except for Monsieur Dargent's visits to Belgium, for Madame had remained on the Continent with their little girl; now she was leaving the thirteen-year-old and coming out with her husband. "When I think of my little girl I am always in sad humor," she told me. Letters were three months reaching Kivu.

The Van Saceghems had a little lad of about ten, home in Belgium. Eighteen months more and they would see him. It was the women who paid colonization's highest price. The life had attracted the men, and all the officials that I met were a splendid class of men, most of them former officers in the war, responsible, discerning, keenly interested in their work. The women were the loyal wives who followed and made a home in the wilderness. I never met one for whom the wilderness itself seemed to have the slightest attraction; I expect they thought we Americans were mad to come so far and go through with so much for the sake of ranging the jungles. When the Belgian women went from post to post they were carried along the beaten ways in a chair by four natives, and they went in white frocks and white shod. They were very domestic; and their housekeeping was infinitely more painstaking even than on the prudent Continent, for here the precious supplies were kept under lock and key and unlocked for each meal.

Madame Van Saceghem got more out of the African life than the others She had a garden radiant with roses brought out from Belgium, which bloomed as roses bloom in Southern California; and she had innumerable pets. We could hardly keep Alice away from this treasure house of little rabbits and chickens and ducklings and goats. Madame had a pet antelope, Bichette, which she fed out of a bottle and often she came to call with the antelope bounding ahead of her, and a golden-crested crane flying overhead. These Kavirondo cranes were birds of beautiful plumage, with black, velvetlike heads from which springs the golden crest. I remember one night when the great crane flapped about a tree in camp while she stood chatting, and the little dyker antelope, enlivened by the evening breeze, dashed about in playful circles, skimming bushes and feigning flight. One could imagine how a band of the little wild things would romp. . . .

Dr. Van Saceghem's work was extraordinarily varied; his real purpose was to investigate the animal diseases, but of course the natives wore a path to his door. While we were there the favorite wife of the chief across the lake brought her sick child to be cured. The wife was a pretty young thing wrapped in scarlet cloth and so heavily weighted with beads and anklets that she would have sunk like a stone in the lake; she never walked more than two steps but rode around in triumph in her litter. . . . She made a social call on us of smiles and silences, for neither of us could make herself understood. Her ornaments were all modern trader's stuff; the tiny white seed pearl beads that were the vogue in Ruanda and the small gold ones. . . . It really is little use to bring beads for trading into the interior, because each place has its own style and fashion and nothing else is of interest. These natives all understood the use of francs perfectly; salt and white American cloth were the only commodities for which they wanted to barter. We bought a little cloth for our own use at Kissenyi of an Arab trader and paid thirty francs for a piece of three yards. We had imagined that we should find a great deal of ivory in the Congo. It was in fact remarkably scarce. We saw very few natives who wore ivory ornaments; this young queen, for instance, had none. The Arab traders, that came in after Stanley had made his way down the Congo, had combed the country and for years poachers and traders had drained the ancient supplies. On the Lualaba we saw girls wearing bracelets of a grain that appeared so like ivory as almost to deceive an expert; a lighted match to the scrapings proved that they were celluloid, part of some trader's supplies. At Katana many of the children had ivory bracelets. They came from the mountains, where there were elephants, and where the natives undoubtedly failed to observe the Belgian order to bring in the ivory of any dead elephant. I saw only two chiefs wearing the old royal bracelets—huge ivory affairs rising to a point. The first time was at Usumbura market, and I was then too politely reserved in my dealings with natives to do more than take a photograph of him; the other time was on the way from Tanganyika, when we camped by the village of Kisaci, a Watussi overlord.

Ruanda is inhabited by two tribes—the Wahuti and the Watussi. The Wahuti are the original inhabitants and they were conquered by the Watussi, a tribe that came down from the north at some dim date, probably during the same migrations of races that brought the Masai down into Central Africa. They are ascribed to Abyssinia, and they have a distinctly foreign air, tall, lean giants of men, often seven feet tall and so thin they appear emaciated, with oval faces and finely finished features. They are the overlords to whom the lands belong; the Wahuti do all the work. The Watussi are not Mohammedans—they have an interesting religion

Photograph by Père Provoost

WATUSSI FAMILY BRINGING MILK AND BUTTER

[page 90]

Photograph oy Père Provoost

WATUSSI GIRLS MAKING BUTTER BY SHAKING CREAM IN A GOURD

[page 90]

White Father's Mission at Nyunde

[page 93]

Cathedral in Erection at Nyunde

[page 93]

in which some analogies to the mediator of Christianity have been traced by one of the White Fathers; but their women maintain a Mohammedan-like seclusion. No one sees a Watussi woman. Musinga is the King over all Ruanda; he maintains a court at Nyanza that is probably the last barbaric court of any importance.

An ancient superstition holds him a prisoner at court, for the medicine men have predicted that if the king crosses a certain mountain he will die and if he passes a certain river he will lose his sight. As a matter of fact neither the king nor the medicine men have any faith in this, for the king did visit Goma, a few miles from Kissenyi, just a little time before we came, though he maintained a strict incognito. The superstition is a convenient way to keep the king within bounds, for if he went traveling and visiting his various chiefs he could take whatever his fancy pleased; as it is, the chiefs go to him, bringing their own selection of gifts. Musinga is fabulously wealthy; every cow and goat and chicken in his kingdom is the king's; and his court has a lavish splendor. He heard of automobiles and sent for two upholstered in red velvet that were brought to him untold miles over the mountains on the heads of carriers and now sit in state at his palace door.

Kisaci, he of the bracelet, was one of the Watussi, a lean old potentate with a scarlet blanket, with a royal assurance and an untiring interest in our camp. He evidently felt very civilized and liked to air a box of safety matches to light his pipe. As he smoked he used to beckon to his *aide de camp* or prime minister or whoever it was, and when the man came, Kisaci leaned his elbow on his shoulder and draped himself against this support for the next two hours.

I noticed at once that he wore a perfectly stunning ivory armlet of the royal shape, and asked to buy it but he firmly declined to sell. It had belonged to untold grandfathers and was worth two wives, but he would not sell it, no, not for the price of two wives. "But for the *toto,*" said I and produced Alice, the Sultana. She was the first white child of course that he had seen, and he succumbed. He consented to give his bracelet to the little one—receiving, however, the francs I had offered—and he made quite a speech through an interpreter, telling her always to remember that the great chief Kisaci of the Watussi had given his royal bracelet to the little white child.

At Kissenyi we saw a few ivory bracelets and were able to obtain most of them through the energy of Madame Van Saceghem, and her own generosity presented us with others. Here, too, we received some of the beautiful baskets made by the wives of the Watussi chiefs which had been given to the administrators as the chief's gifts. Madame Keyser at Changugu, had given us her own. They were round and high with a peaked top, woven with intricate fineness, with bizarre jagged black lines of patterns.

While waiting at Kissenyi for the porters to come in and take us up the gorilla mountains, and for provisions of dried beans to sustain the porters upon the expedition, Herbert and Martha and I made a day's excursion into the mountains east of Kissenyi to the mission of Nyunde, from which we were drawing our generous supply of vegetables and fruit. The way wound along rich slopes checkered with fertile fields where the women were working—it is the women and sometimes the children who do the field work among natives.

Almost every woman had a baby tied in the goat skin on her back, and as the women would hoe away the little head would bump and bump against the maternal back or else drop in sleep at an angle harrowing to our sympathies. Only twice during those months in Africa did I see a child laid in the bushes while the mother worked. I suppose the danger from wild beasts, leopards especially, occasioned the custom. The father holds the mother responsible, so she carts even a good-sized youngster with her. She has a stoic endurance, but she has also a stoic indifference. I have seen women gossiping and laughing while a baby was shrieking itself hoarse. The infant mortality is tremendously high; a great deal of it must be due to the native fashion of stuffing a child. When the baby is a few months old the mother begins to feed it a preparation of coarse meal; she forces it down in much the fashion that the Strassburg goose is fed; filling her hand with the stuff she pushes it into the baby's mouth and holds the little nose until the gagging child swallows. Often the baby has spasms of coughing ending literally in convulsions; when it is exhausted, she begins again. Custom is a wonderful thing. A native mother would regard herself as a moral derelict if she shirked this maternal duty.

The Mission at Nyunde has a superb site, crowning a mountain top; the outlook over the amazingly fertile valleys and mountains reminded one of scenes in the

Alps. For twenty years the Mission has been there; the Father Superior had been there for thirteen years. Before the war, he told me, there were five thousand black Christians in the vicinity, now famine has reduced the number one-half. I do not know exactly what he meant by a black Christian. The Catholic Encyclopædia for 1911 reported that there were 4823 blacks in the Congo baptized by the White Fathers and 18,797 catechumens. The fathers grew the vegetables and fruits of Europe in their gardens, experimented constantly with grains and seeds, cultivated coffee and tobacco, and made thousands of cigars. They had taught the natives brick making and a large brick cathedral was in course of construction.

At the foot of the hill were the buildings of the White Sisters. One of the four was from Canada, but being a French Canadian she spoke no English. Another was German. The order is an international one, founded by Cardinal Lavigerie. French is the universal language.

We lunched with the Fathers, then visited the Sisters and photographed them among their roses; they had the sweet and tranquil faces of women who have yielded every personal expectation. They were disappointed that I had not brought Alice. "To have seen a white child!" said one of them.

Another interesting excursion that we made about Kivu was to the lava flow at the end of the lake. The north end of Kivu is the fringe of that volcanic field of which the eight distant separate volcanic peaks, the

WHITE SISTERS AT NYUNDE

[page 94]

MT. CHANINAGONGO

[page 94]

Native Fishing in Kivu

[page 97]

Flow of Lava across Lake Kivu

[page 97]

M'fumbiro, dispute with the Ruenzori the legendary title of Mountains of the Moon.

Certainly these mountains are the farthest source of the Nile, for on the southern slopes of the Sabinio, M'gahinga, and M'havura volcanoes, the Nyjawaronga River takes its rise, drains first to the south, becoming known as the Kagera River, then flows north and finally empties into the Victoria Nyanza. Also the Ruchuru River rises not ten miles north of Sabinio and flows into Lake Edward, thus uniting with the Nile system.

It was the three highest peaks of these Moon Mountains that were our objective — Mikeno, Karisimbi, Visoke. The two others were Chaninagonga and Nyamlagira, whose craters still guarded that deep fire which had flung them into being. East and west of this group run the bastionlike mountain ranges that shut in the length of the Kivu-Edward Rift valley, and these mountains are neither of volcanic origin nor connected in any way with the volcanoes which rise from the floor of the valley itself.

The north shores of Kivu are of lava rock, giving an utterly different aspect to the lake than that of the green grass hills of the southern setting. This M'fumbiro region is probably the area of some of the most recent volcanic activity in the world, but practically nothing of its history is known and but very little yet reconstructed. We are indebted to the fortunate chance of a visit of Sir Alfred Sharpe, formerly governor of Nyasaland, to the shores of Kivu in December, 1912, for the account of an eye-witness to a violent eruption which altered the entire northern extremity of the lake.

From a fissure in the earth about two miles north of the shore, a fissure that speedily developed into a small cone from the rocks and ash and cinder flung up, there had come such a tremendous outpouring of molten lava that the north end of Kivu, already an inlet in character through the irregularity of its shore line, was almost shut off. Entire tracts of the country were destroyed and hundreds of natives killed; it was the Last Days of Pompeii for many a luckless village. The boiling lava poured into the lake in such quantities that the water seethed for twelve miles out and many fleeing natives who escaped the fiery rocks and engulfing lava by launching their hollow log canoes were capsized into the scalding waters and fairly boiled to death. Walikali, a hundred and fifty miles west, was covered with ash and the sound of the eruptions was heard as far as Pili-Pili, two hundred miles away.

Everything had been quiet now for ten years, so we felt it fairly safe to visit the devastations. By paying for the precious gasoline we obtained the privilege of using the governmental launch for the day, and with the Van Saceghems and Monsieur Jungas, the *Procureur du Roi,* a judge of the highest court in Stanleyville, who was making his rounds with his assistant, we made an excursion to it. It took about four hours by boat to reach the long tongue of lava reaching out from the rocky entrance to the Kabinia Inlet, as the north end of the lake is called; we managed after considerable difficulty to find a place on the rugged shore where we could make a landing. Then for some time we explored the extraordinary field of lava, solidified in streams and swirls like frozen black whipped cream over which the green of lichen was creeping, and in whose crevices the first long grass and wild flowers were finding root, while wild morning glories spread out like an old-fashioned floral carpet before us.

On our return we saw through the mists of the afternoon rain the shrouded outlines of Nyamlagira, whose fire made the ruddy glow each night on our horizon, and the cloud-wrapped peaks of Mikeno and Karisimbi, our distant gorilla heights.

CHAPTER VII

THE GORILLA TRAIL

Into the Gorilla Forests of Mount Mikeno

It was November the ninth that my husband and I and Martha Miller and Priscilla Hall and little Alice and a string of one hundred and seventy porters left our camp at Kissenyi and started on the three-day march to the Mission of the White Fathers on the slopes of Mikeno, from which we expected to make our ascent after gorillas.

Mr. Akeley, with thirty porters, had left ten days before and was now encamped on the heights from which he sent down constant news by runner. His reports of the conditions under which he had been able to see gorillas were full of disheartening difficulties, but the fact remained that he had been able to see them and obtain some desired specimens for his group, so Miss Miller and I were undiscouraged. We were prepared to climb and crawl and freeze, but we were going to see gorillas. We had not come up the long trail from the Cape to be turned back by any hardships now.

The reports which we obtained from Mr. Barns of the conditions on the other side of the mountain were more heartening. Mr. Barns had come down one side as Mr. Akeley had gone up the other. He passed within a few hours' journey of Kissenyi and Herbert and I went out to meet him. Mr. and Mrs. Barns were encamped near Nyunde drying the gorilla skins, and from there he was going on after some monkeys and chimpanzees. These English people were a very interesting couple; he was a tall, spare, black-haired man, about six feet two, who had tired of a Rhodesian ranch and had turned his hunting to collecting; his wife was young and vivacious, with bobbed, curly hair; for eleven years she had been in Africa with him, never hunting herself, but sharing the dangers of the life.

Some years before he had shot a gorilla that was mounted by Ward for the Rothschild Museum; and now he had come for gorillas for the British Museum and had secured three for a family group. They were beautiful skins, the female and young one very black and shaggy like bears, the male with the distinguishing silver-haired back. This male had an even larger head than the one Mr. Barns had shot two years before; it stood five feet three and three-quarter inches high, measuring seven feet six and a half inches from the ground to the finger tip. From finger to finger it measured seven feet six inches and the chest measured fifty-seven inches. The chest measure of his former gorilla was sixty-three inches.

He had come up with this old male in an open space on the edge of a precipitous ravine high on Mikeno, and the gorilla had started towards him, roaring and beating his breast. This was quite like Du Chaillu's description of their behavior and I asked Mr. Barns if he considered this a charge on sight from a gorilla, and he said he couldn't say—the beast might have been acting from surprise or alarm and trying to intimidate him. But he certainly started towards Mr. Barns. In the brute's excitement his progress on his hind legs was unstable, and after coming a little distance he stumbled and fell down the clifflike ravine he had been skirting. The fall knocked the fight out of him. Mr. Barns went down the ravine after him and followed through the jungle for an hour and a half before coming up with him. Then he killed him with a single shot.

The perfection of the modern weapon does not give the gorilla the chance he had in the old muzzle-loading days. In almost every story I had heard of recent gorilla hunting the gorilla had been taken by surprise and shot at once. If wounded he would very naturally turn on his attacker; whether he would charge on sight, if unattacked, was something we did not know.

Mr. Akeley believed he would not. My husband and I had no pretensions to a conviction of any sort, but we were going to try to meet a gorilla on his native heath and find out what he would do about it.

Mr. Barns had one bit of evidence as to the behavior of the female of the species. He told of crawling half an hour on his hands and knees through the bamboos and then coming suddenly face to face with an old lady gorilla. She was as much surprised as he was and evidently had no desire to continue the chance acquaintance, for she made off at all speed. He let her escape, for he was on the hunt for a male.

After seeing the gorilla skins, the gorillas themselves seemed more real and less legendary, but there were so many difficulties in the way of discovering them, and it was all such a matter of chance in spite of the hardest

THE WOMAN WITH A HOE
[page 101]

OUR POSTMAN
[page 101]

The White Fathers at Lulenga Mission

Left to Right—The Frère Hyacinth, the Père Provoost, the Père Vander Handt, the Père Superieur Von Hoef

[page 103]

Lulenga Valley. Mission White Fathers

[page 103]

sort of pursuit, that we did not feel in the least sure whether we three amateurs would have any real luck or not. But we meant to stick it out until we saw something of them.

We marched through very beautiful country the first day out from Kissenyi, with magnificent views of the volcanoes. Monsieur Wera, the *chef-de-poste* at Kissenyi, had lent me his donkey for that first day's ride, much to Alice's joy, who confiscated it half the time. Priscilla and Martha alternated walking with the *machila* or hammock. The camp that night was at Kibati, with a marvelous view of the volcanoes and of Kivu.

At Kibati a little enclosure of elephant grass shut in eleven graves of officers dead in the Great War, pathetic graves marked with wooden crosses rudely carved, weighted down with stones to keep the beasts away, encircled with the roses and geraniums from Europe planted by some kindly hand, and shaded by lemon trees and plums. We walked among them that afternoon. . . .

CI GÎT LE CAPTAINE I. J. DE FOIN
TOMBÉ GLORIEUSEMENT AU TSAND JARNWE
27-11-1915

LE SOUS-OFFICIER J. L. DE VOLDER
TUÉ À L'ENNEMI EN PORTANT SECOURS A SON CHÊF
À LA MEMÔIRE DU SOUS-OFFICIER ALFRED DUPUIS
MORT EN BRAVE

J. CORNESSE
COMMANDANT—MORT POUR LA PATRIE

How little those men expected, playing as short-socked little boys in some trim Belgian garden, to come to their death on the African mountains . . . to lie weighted with lava rock to keep the hyenas away. *Mort pour la patrie!* . . . What forgotten courage and despair had played their hour in those trenches that zigzagged the mountain sides, overrun already with oblivious green.

Our second day's march brought us to Bubonde, the territory of Burunga, an independent old chief or sultan, who had an evil name among the missionaries for robbing caravans. The disorders of war times had given him quite a field for activities, but now he was reduced to cheating his neighbors in the matter of cows. It was rumored that he was to have a white man for a son-in-law; that a German planter who had exhausted white credit and forbearance, who had tried everything in fact but work, after much lolling in an easy chair, drinking pombé, and overseeing his workers through opera glasses, had decided to "go native," and, for the sake of Burunga's cows and shambas, was taking one of his daughters in marriage. Usually a man pays the father for a wife. The planter was evidently getting his with a dower. . . . That was rumor. I tell the tale as it was told to me.

Leaving Bubonde the third morning we went through thick forest, up hill and down, on paths that were harsh lava rock. Lovely pink flowers, like starry orchids but growing on silvery green bushes, edged our way. Two hours brought us up the last steep climb and to an opening on the lower slopes of Mikeno, where the thatched roofs of the Lulenga Mission showed among the shambas.

Three Fathers and a Brother occupy the Mission which has been in that vicinity twelve years; the Père Superieur von Hoef, the Père Provoost, the Père Van der Handt and the Brother Hyacinth. They hospitably installed us in the house prepared seven years before for the White Sisters, who were now expected each week; I said something about the seven years' wait for Rachel; kindly they overlooked it. It was a long, low, mud house, whitewashed outside and in, with dirt floors and thatched roof. The windows had wooden shutters; there were no screens needed—any glass, of course, did not exist. At Kissenyi, Monsieur Van Saceghem's house had boasted actual window panes, some of glass from photographic plates, and some from the celluloid of old films. Little schoolhouses, three-sided mud affairs, had been built back of the Sisters' house; tilled fields stretched on the steep slopes before; roses were planted about and edged the straight way north to the Mission of the Fathers and the Church.

It seemed heavenly to our feminine souls to have a room and a table again; we could imagine nothing more luxurious except a bureau with drawers. Really to appreciate a dresser drawer one should have lived in a tent, with all one's belongings incarcerated in barrel-like bags, and the desired belonging always eluding one's blind groping.

Before the house stretched a marvelous view of the Rift Valley—green hills reaching down to wide-spreading lava-fields from which rose the small craters and cones of the valley floor, and across from us the shadowy slopes of the mountains that guarded the west. South of them rose the beautiful outlines of Nyamlagira and Chaninagongo, above whose craters hung a ruddy smoke. Chaninagongo smokes rather lazily, but Nyamlagira has been steadily active since its last eruption seven years ago, and at night the glow from that crater was a flame in the sky.

Lulenga Mission has been the lodestar of gorilla-seekers ever since the discovery of true gorillas in the heart of equatorial Africa. Gorillas had always been associated with low, west coast jungles of the Gaboon, but twenty years ago an Englishman, Quentin Grogan, on his famous two-year walk from Cape to Cairo found the skeleton of a true gorilla in the equatorial mountains.

Later, occasional rumors of great apes up in the bamboos were carried by the natives to the white men, but the apes were supposed to be chimpanzees whose presence was already known.

Then about thirteen years ago an Austrian, Grauer, passed through Nairobi with some gorilla skins he had obtained in the mountains of what was then the western edge of German East. The first gorilla of which the Mission had any record was that shot by Count Pauwels in the commencement of 1913. At the end of 1913 Count Arhenius made a successful hunt in the bamboos. Eight years later he returned with Prince William of Sweden on an expedition which was just leaving the Congo as we came into it. The Prince had made his camp here at the Mission and shot his first gorilla here in the forest just an hour from the camp. The other members of his party, even the soldiers, had killed others, so altogether fourteen were slaughtered.

There were moments when we wondered anxiously if there were any gorillas left for us, anything but lone widows and undergrown youths. Of course a gorilla was a gorilla, but it was the grown male who had given the legend of ferocity to his race.

I had brought with me Du Chaillu's description of his first sight of one.

> The underbrush swayed slightly just ahead, and presently before us stood an immense male gorilla. He had gone through the jungle on all fours; but when he saw our party he erected himself and looked us boldly in the face. He stood about a dozen yards from us and was a sight I think never to forget. Nearly six feet high (he proved two inches shorter) with immense body, huge chest, and great muscular arms, with fiercely glaring large deep gray eyes, and a hellish expression of face, which seemed to me like some nightmare vision; thus stood before us the king of the African forests. . . .
>
> His eyes began to flash a fiercer fire as we stood motionless on the defensive, and the crest of short hair which stands on his forehead began to twitch rapidly up and down, while his powerful fangs were shown as he again sent forth a thunderous roar. And now truly he reminded me of nothing but some hellish dream creature—a being of that hideous order, half beast, half man, which we find pictured by old artists in representations of the infernal regions.

In the hope of meeting the gentleman of this description we started off up the mountain at six o'clock the morning after our arrival at the Mission.

It was a lovely morning, sweet and fresh, with that feeling of spring that mornings have in Africa. We were mounted on a mule and donkey belonging to the White Fathers, with guides running ahead and boys coming after with guns and water bottles and camera. We rode nearly an hour up the steep foothills of Mikeno.

Back of us the brown roofs of the Mission grew smaller and smaller, the waving fronds of the banana plantations merged into a sea of green darker and more glistening than the tender green of the fresh grass about them. The great plain of lava on the valley floor stretched wider and wider as we mounted higher, and across it the miles and miles of mountain peaks were blue against the sky.

Ahead of us the sharp, craggy peak of Mikeno stood out in bold relief, with glistening clouds floating below it, shining with the first sunlight. It was a stiff climb and our hearts ached for the puffing mules struggling up the slippery, narrow path of mud, but we let our hearts ache and conserved our legs.

At seven we were on the edge of the forest, and started on foot up a narrow path, tilted at a violent incline, a path like a greased chute of mud. It might have been a rush of water after a heavy rain, but now the mud was a smooth spread, sometimes a slippery smear over rocks, sometimes a slough of incredible depth.

The trees shut us in, the vines netted us like basket work. The guides climbed ahead, my husband after, his gunbearer behind, then I followed with my gun boy behind me. We kept our eyes sharply on the path and suddenly I saw in it a print, perfect in the soft mud.

It was a hand print, the fingers doubled under, showing the marks of four knuckles and a thumb. A little ahead were the outspread toe prints, where Herbert and guides were pausing as I saw the hand marks.

They were gorilla prints, freshly made. The guide declared them "Kubwa, kubwa" (Big, big), which was stirring. The gorilla had been walking along the path, helping himself by his low hanging fists. He avoided the deep mud, keeping to one side of the chute where the ground was what I should call mushy, but mere mush had no terrors for him.

We followed with a feeling of tremendous exhilaration. It was the actual mark of the great beast we had come so far to see; he was there somewhere ahead of us, hidden in a turning of the green thicket—any moment a parting of the leaves might show us his black, twitching face and sparkling eyes.

We ordered the camera boy to keep close and we kept the gun boys extremely close. We had been cautioned not to trust our guns to these carriers who were not gun boys in any sense of the word, and were quite likely to cast the guns away and run at a critical moment. However, the climbing was much too hard to do with a gun in our hands and we took our chances.

At every turn we gazed about hopefully, remembering that the Père Van Hoef, the Father Superior at the Mission, when hunting with the prince, had suddenly seen and shot a gorilla in the branches of the tree just by his head, but no ape disclosed itself. The path, however, revealed interesting secrets. Here were antelope tracks; here was the sudden spring of a leopard, half dried in the sun. Suddenly it ceased to be a path and became a series of artesian wells. These were elephant tracks like bottomless pits, freshly made, with water still slipping into them.

They were difficult to negotiate if one tried to step across and balance on the mud ridge between, but we clung to the bushes at the side and got on. Finally the tracks crossed off to the left where we saw tusk marks on a tree trunk.

It made me remember poignantly that the large gun was at the Mission. Herbert and I were each carrying a Springfield. We had nothing heavier for a charging elephant. I remembered it again, even more poignantly when a tiny sound held us motionless. It was a snapping and tearing of twigs.

The guide crept closer. His low-breathed "Tembo" (elephant) was almost inaudible in his anxiety not to be overheard. I stared hard at the bamboo screen, but it was impenetrable. I hadn't the faintest idea how far away that elephant was breakfasting, but I had no desire to find out.

We were then entering the bamboos, a forest of tall, slender stalks and delicate leaves, all netted and interwoven with vines. It was colder here, so gray that the sun seemed to be under a cloud. We pressed harder on the trail, trying to catch up with that gorilla, and suddenly came into a little clearing, sun flooded, filled with delicious young growth, a heavenly place for a picture of a gorilla at breakfast.

No applicant appeared. Instead the spoor vanished. We paused to let our guides munch some hard, berry-like grain and smoke their black clay pipes, while we ate chocolate and crackers; then we urged them to fresh effort. But the trail was lost.

They led us at last to one of the innumerable little trails that led out from the morass through the creeping vines, and on we went into ever dimmer and more impenetrable solitudes. Hanging vines hung down like tapestries and a network of them veiled the undergrowth. The guides hacked away with their sickles and we wormed our way along, often forced to crawl on all fours through some bad bits. This went on for hours. The guides had apparently given up all hope of a gorilla, but were going to earn their francs by exercising us.

We kept on doggedly till at last, discouraged by our persistence, they united in calling it a day and began to slide down the ravinelike sides. We got back after seven hours, heartily tired, having accomplished nothing of the morning's hopes but the sight of that gorilla trail, yet we had spent a thoroughly fascinating day.

CHAPTER VIII

THE BIG GORILLA OF KARISIMBI

THE GORILLA HUNT IN WHICH HERBERT GETS THE BIG MALE OF KARISIMBI

WE had intended starting again on Monday, but a runner from Mr. Akeley caused us to change our plans. He wrote that he was ill, that he had "broken something." So, on Monday, Herbert and Martha Miller and I and sixty porters started up the mountain to his camp, leaving Alice and Priscilla Hall in the White Sisters' house. It was a great comfort to have them there, where the untiring, kind Fathers did everything possible for them.

Mr. Akeley had written that we had better take two days to the trip as only veterans might make it in one, but we felt decidedly veteranish by now, and, as his letter made us anxious, determined to get through at all cost. We had not gone more than three hours before we came up with the porters opening their loads with unusual alacrity; the cook was busy spreading out his magnificent red mattress which constituted an entire porter's load.

This was in a damp glade on the mountainside, and the march we had made was not a day's work, so I—being ahead—told them to go on with all the vehemence and Swahili I possessed. *"Pana mazuri hapa"* (not good here) was repeated vigorously until they got up and hoisted their loads upon their heads. They didn't really hope to put it over, the headmen had been told at the Mission they were to go all the way, but they considered it decidedly worth trying.

Later I was to hear that *Pana mazuri* flung back at me by a half a hundred of them as they slipped and sloshed and scrambled up that mountain's sides.

We had thought we could go up in six hours. It took us nine. The last six hours were a steady, interminable climb, up through the forest, into the bamboos, through the bamboos into a higher forest again. The path was the same sort of mud chute that Herbert and I had climbed before, and we had to cling to the trees at the side for leverage.

I understood then why soldiers at the Front had thrown away rations, water, ammunition. Sometimes every step seemed literally the last possible effort. The altitude had its effect, of course, in conjunction with the continual struggle.

There were times, about the sixth hour, when we found cheer in song, peculiarly suitable songs such as "There's a long, long trail a-winding," and "Smile, smile, smile," but after that our breath gave out and we saved it for such valuable speech as "Rest here—we can take the day to it."

But our spirits did not flag. When Herbert, following our steps and watching us, chuckled at the load of mud that went up and down with each foot and announced that we wouldn't do for fairy-footed partners at a dance, we looked at his own weighted feet and assured him of his complete unsuitability.

At intervals he cheered us on by telling us we were going to be the only women in the world who had seen wild gorillas. We retorted that we hoped they'd appreciate the trouble we were taking and if a wild gorilla would only appear and perform that much advertised act of carrying women off we wouldn't offer any resistance.

The end of the ninth hour we reached the camp, and found Mr. Akeley looking as if years instead of days had intervened. He was very worn; he had done the work of ten men under particularly trying conditions; he had started with a fever, infected by jiggers which he had not been able to extract; he had killed his gorillas in most inaccessible places where the natives had balked at following; he had skinned and skeletonized and dissected without rest, and now energy and appetite had deserted him. What was broken, he said, was his vigor. We felt troubled when we first saw him, but a good dinner, an incentive toward appetite, began to make him feel better.

The camp was high on Mikeno, the mountain's citadel-like crags above, a world of forest and valleys and mountains at its feet, with clouds floating up the chasms and stealing among the trees. There was only a tiny clearing for the tents with the porter's huts of grass tucked in behind; the gorilla skeletons were hanging from poles in grisly sociability, while from the tent of Mr. Akeley hung a small, mummified figure, a skinned

Our Objective—the Gorilla Triangle. Mt. Mikeno in Foreground, Mt. Karisimbi at Right and Mt. Visoke, Left

[page 112]

Gorilla Camp on Mt. Mikeno

[page 112]

WHERE GORILLAS LIVE—THE FAIRY FORESTS OF KARISIMBI
[page 115]

THE BIG GORILLA OF KARISIMBI SHOT BY HERBERT E. BRADLEY
[page 115]

and dried two-year-old gorilla whom we christened "Clarence."

Beside securing his specimens, Mr. Akeley had fulfilled the hope which had been only a dream of the expedition: he had taken motion pictures of wild gorillas—a mother and two little ones—something that had never before been done.

The next day we spent quietly in camp. Wednesday, we left our boy Kiani in charge of things there and started off to make camp higher upon the ridge between Mikeno and Karisimbi.

It was a two-hours' climb and we camped in a glade full of flowers, wild carrot, and buttercup, with a marsh before us, reaching to the forested side of Karisimbi.

Balmy as that glade was in the noonday sun, the night was a revelation of Arctic chill. Our preparations for bed were elaborate, but even so, the temperature surprised us, and what it did to our scantily clad porters, huddling blanketless in grass huts about their smoky little fires, we could more than surmise from their conversation during the night and morning.

It did not actually freeze. It was like northern Wisconsin in late October, when your breath hangs in a cold cloud in the air before you. We had no camp fire, for the wood was wet and smoky, and we had only an iron bucket of wood coals to warm the tents.

Thursday we started out for the gorilla pictures which it was the hope to get before adding any other specimen to the group. The guides led us up the Karisimbi slopes—only slopes is too gentle a word—and we climbed and climbed.

We were in a fairy forest, trees gray with lichen and green with cushioning moss, trees dripping with ferns and garlanded with vines. When the sun shone through that forest the moss gleamed in golden richness. There were trees with sharp, down-pointed leaves, with a russet glow from the leaf stalk, that hung like a jeweled filigree against the tropic blue of the sky. There were clouds of pink, orchidlike flowers, that were not parasites like orchids, but grew on silver green bushes, and everywhere were snowy reaches of wild carrot and wild parsnip and the familiar pungency of crushed catnip.

There are no words to describe that forest. Pictures can give but faint clues. It was a magic spot. Arthur Rackham has dreamed some of its moods, some of its wizard trees with long curved arms, its crooked, bending groves, like magicians in flight; but its color, its delicacy, the infinite fragility of its moods, the seduction of every line, the subtle revelation of its lights are beyond dreams.

We found no gorilla that day. We found raspberries instead, enormously large and extremely green, and we also found fresh traces of buffalo. The guides were eager about both. They consumed the berries and pointed out the traces of "yama" (meat) eagerly, so we concluded that they had gone upon a replenishing expedition. After five and a half hours of thorough exercise we went back to camp, having cleaned up that section and found no gorilla trails.

Friday opened with glorious sunshine and an ultimatum from the guides. They were going. The cold nights prevailed over the passion for francs. They had enough now anyway for several wives and a long life-time of ease. However, they were prevailed upon to wait one more day and we started forth in haste before they changed their minds.

This time we took another trail up the Karisimbi heights, with ever more and more glorious views as we climbed. At last Mr. Akeley halted. "This is the most beautiful place in the world and I am going to photograph it," he announced with a certain defiance, knowing the guides viewed any dallying with the cameras with distaste. They understood a gun; the camera was, to them, resultless.

But he did not; as he poised his machine, the men pointed. On the slopes to the left the bushes were waving, giving a glimpse of something like a black bear.

Hurriedly we marshaled in line and scrambled up the trail, then in and out the trees and bushes, Herbert and Mr. Akeley first, then Martha and I, our gun boys, though relieved of our guns, hurrying excitedly after us. We went under a hollow tree feet first and emerged on the other side with a clear view of the slopes before us. There, on the steep mountainside stood a gigantic creature, black and shaggy. My first impression was of shoulders — incredible shoulders — huge, uncouth, slouching shoulders. His side was toward us and his back was silver gray. We were seeing at last the great beast we had come so far to see—a male gorilla in his savage haunts.

It seemed an eternity before my husband fired. I suppose it was only an instant or two. The roar of the gun sounded as unreal in the silence as the sight of the gorilla. Immediately the gorilla went crashing down into the welter of vegetation. We thought him dead and raced down towards him after Herbert, but we then found he had made off, leaving a trail of crushed greenery and blood. For a few moments the waving bushes gave us the only clue, then he emerged on the slopes above and looked back over his shaggy shoulder as the gun crashed again, as if trying to comprehend this sudden assault upon his solitudes. I shall never forget the humanness of that black, upturned face.

Then he went plunging down the slope, passing near Herbert, who put in a finishing shot. The great body struck against a tree and lay still. There had been no sound from him, no bark or roar. He had shown no instinct of fight, nothing but the rush of a wounded beast to escape.

We found him dead against the tree, face down, a huge, shaggy, primeval thing, like something summoned out of the vanished ages. And the scene in which he lay had a beauty that was like nothing earthly.

From that high place, whose forested slopes swept down, down, like a green flood to the distant valleys and the blue sheen of Kivu, we looked out across to the purple heights of Chaninagongo and Nyamlagira, crested with clouds that were golden with sun and rose with volcanic fires. To our right, sharply silhouetted against the distant azure and amethyst, stretched the superb slope of Mikeno edged with delicate little trees, exquisite miniatures relieving that long line, that went up, up, to the citadel crag of the top, glowing with umbers and emerald moss.

The gorilla proved a huge gray-backed male. When he was tugged and propped upright I shall never forget the impression he made. The great girth, the thickness and length of arm, the astounding shoulders made him a giant.

His face was ferocious only when the mouth was open. The normal expression was of a curiously mild and patriarchal dignity. Without being sentimental you could see in that face a gleam of patient and tragic surmise, as if the old fellow had a prescience that something was happening in the world against which his strength was of no avail—as if he knew the security of his high place was gone.

For generations he had lived without fear. He preyed upon no one for his food; he ate the wild carrot and fresh greens, disturbing no one and disturbed by none of his world. He could have crushed a lion or strangled it, and an elephant, if gripped by the trunk, would have no thought but of escape. He had been indeed the King of the African forests.

We took measurements and found his height to be five feet seven and a half inches; the reach from his upraised hand to the ground eight feet and two inches, and from hand to hand seven feet eight and a half inches. His chest was between sixty-two and sixty-three inches. He was, we feel sure, the big bull of Karisimbi, of which we had heard. This bull had been shot at before and we found an old wound in the hip, which had given a decided curvature of the spine, shortening the height.

Looking at his great arm and curving fingers, the fist as big as a man's head, I could understand how unwary hunters in the old days had been scooped out like soft-shell crabs.

All that day the men worked on the gorilla, for Mr. Akeley preserved everything for museum and medical records. They paused often to photograph the changing clouds and mountains.

It was a marvelous day! The sheer beauty of it was a spell, and the presence of this great gorilla made it seem like a page from the very beginning of time.

CHAPTER IX

A GORILLA BAND

GORILLAS IN GROUPS; A MOTION PICTURE HUNT

THE day after the big gorilla of Karisimbi was shot the guides all left us. They refused to endure the cold and hardships of the hunting for another day. They were really a fine lot of men, the Wahunde, belonging to the Sultan Burunga, whose shambas were perhaps a quarter of the way up the mountain, and they had been working hard during the day and nearly perishing with cold at night.

Sculpturally they were magnificent specimens, and their swinging goat skins were suitably picturesque for their dark, muscular bodies, but a girding goat skin is small protection against cold by night and against nettles by day, and those wonderful forests of the mountain heights were stingingly alive with nettles.

So we could not blame the guides for feeling that they had had enough of it, although we saw them go with regret, and the boys and porters eyed their departure with yearning envy.

Then the porters gathered and frenziedly announced that they, too, were going. They had business, urgent business, elsewhere. They came from villages beyond Kissenyi, and they had imperative reasons to return to them. One man declared he had to accompany his chief on an expedition to Musinga's court, and another remembered suddenly that a neighbor had threatened to kill his son, and he wanted to go right away and exterminate that neighbor. Others did not produce any reasons, but stated their intentions.

But they were in no position to issue ultimatums. We could not control the guides but the porters had enlisted, as it were, for the duration of the war, and would not be paid until we came down from the hunt. With all their pay coming to them and the extra backsheesh they were to receive for staying, we felt fairly sure that they would not desert.

But we did not feel that we could keep them another night in that high altitude, on the exposed saddle between Karisimbi and Mikeno. They tried to keep themselves warm with fires in their grass huts, but their chief resource was to huddle together like close packed cattle and exercise their lungs. They talked all night—generally about us. You could not wake at any hour without hearing the "Wazungu" (Europeans) getting theirs from some impassioned orator!

And often the hearty roars of laughter told us that the village wag was not overlooking any little thing they found quaint in us.

They had no blankets—nothing but the goat skins and wisps of cloth they wore by day. And one man lost even these. It was the morning the guides left. He came over to us with a wreath of greenery about his waist, like a Russian dancer performing Spring, and amidst the hilarious laughter of his friends he delivered a long, pleading, and utterly unintelligible discourse.

Mr. Akeley Working on Gorilla Skins

[page 120]

Guides Removing Flesh from Gorilla Skeletons

[page 120]

Gorilla Skeletons—and Others

[page 121]

Gorilla Camp—Skeletons and Skins Drying—Little Clarence Hanging in Tent

[page 121]

Being interpreted, it appeared that he had burned his garment in the fire that night and was asking the loan of a dishcloth. Cloth was too precious for generosity. We gave him half a dishcloth. He was quite pleased and abandoned his wreathing buttercups with relief.

We sent the porters on ahead of us with tents and supplies to make camp in our former location in Mikeno about a thousand feet lower and much warmer than this glade, and then we scouted back over Karisimbi where we had found the big one the day before, to try for photographs. We tried for five hours without a glimpse of any living thing. Not even a leopard had visited the kill. We had seen leopard's tracks near our tent, and the boys were in deadly terror of them at night; but they could not be plentiful or they would not have overlooked the gorilla meat which we found lying just where it had been cut from the skeleton.

The natives did not eat the meat. We ourselves had cooked and eaten a little, just for the sake of doing it, and found it perfectly good meat, firm and sweet, but I couldn't get over the family feeling of sampling grand-uncle Africanus!

We came into the Mikeno camp that night feeling thoroughly exercised, as we had a habit of saying. The next day, my husband and I started off alone with the idea of locating a troop for a motion picture if possible, and, if the light should be too gray for the picture, of killing a female for the group.

Having no guides we took our luckless tent boys and porters to carry our guns and plate camera. For over three hours we led them a hard chase through the nettles that burned like fire wherever they touched. At every rest the men scratched vigorously and told each other in tones that were perfectly intelligible, however unknown the words, their opinion of our excursion.

There was one frank and thorny nettle that could be recognized and generally avoided, but there was a soft, plushy-looking plant, seemingly as mild as a nursery tea, that left a wipe of almost invisible bristles on you like a fairy shaving brush, and the touch of it burned a full hour. The only consolation in getting another touch was that you then ceased to occupy yourself with the old one!

Finally, after some hours, we struck an old antelope trail that promised easier going and we wound up and down the slopes on that, but found nothing but the antelope marks. Once we came on a small bush buck, but he did not stay to have his picture taken. Then we passed a gorilla trail leading off to the left, which gave some sign of promise, broken stalks showing where the gorillas had fed.

Judgment told us to follow that trail, but when my husband asked me which way I wanted to go, I abandoned judgment and clung to the easier footing of the antelope path, although it looked too much used here by the natives who came up from Burunga's into these heights for wood, for chances of gorilla. But I voted to keep on with the antelope path until we had found a good spot for lunch and rest, and so we sauntered easily on and then—just to show how luck and judgment differ—we rounded a mountain slope, passed around a fallen rock and tree and looked down into a lovely, open meadow where four gorillas were feeding!

I saw only three at first. There was a huge silver-backed male, slouching along with his strange shamble among the waving green, and two females, recognized by their black backs, a little distance behind him. Their gait is extraordinary. They walk on their hind feet, assisting themselves with their hands, the fingers knuckled under, the thumb outspread, and, as their arms are enormously long, their backs are but slightly bent.

It gave us a tremendous thrill to see those great beasts there in the glade. I felt a rush of exultation at our luck. Three of them! And one of them the very gentleman we had come so far to interview. "Now," I remember thinking excitedly, "now, we are going to *know*——!"

We had frozen at attention the instant we saw them, signaling the porters to keep silent and out of sight; but some sound had already given warning of our approach. The big male looked around straight up at us, rising as he did so.

It was an uncannily human face that he turned up to us, but there was none of Du Chaillu's demon horror about it. I got an impression of a wary interest that did not intend to tolerate any intrusion, but there was not a flash of menace—nothing that the most prudish person could possibly call hellish. He simply conveyed the idea that he had been disturbed by a distinct outsider, and started deliberately away, shambling along through his ancestral meadows towards an arch in the trees leading into glades ahead.

The two females came after him, like big shaggy black bears, except for their peculiar gait. None of them did any barking or roaring and there was nothing to alarm the fourth gorilla, which I saw then for the first time, as she suddenly discovered the others making off without her and came lurching rapidly after them.

I had already pulled out a bit of paper, a chocolate wrapping, from the convenient safari pockets, while Herbert thrust his pencil into my hands, and started a swift scrawl to Mr. Akeley telling him to hurry with the camera as we had come up to three gorillas—and at that moment I crossed out the three and wrote four.

The note was passed back to a porter who was ordered to go, *pacy, pacy,* a short cut to camp to bring Bwana Akeley with his camera here.

It was just twelve-thirty. The sun was out and the light would be good for five hours, and if the gorillas were not hurried or frightened there was every chance of coming up to them.

One after another, not running, but with quickness for all their seeming clumsiness, the four gorillas went on under the trees, looking back constantly in that terribly human way. Then they disappeared.

It was hard not to keep after them. We settled down in a thicket that commanded a view of the meadow, with our boy Aluma, who had been carrying my gun, beside us, and sent the men farther back on the path. While we watched we saw two or three of the gorillas appear in the opening into the glades beyond and cross back and forth, apparently feeding. We could see them

IN THE GORILLA FORESTS OF MT. MIKENO

[page 124]

GORILLA BED AND TRAIL

[page 124]

Gorilla Bed, Overhung by Ferns

[page 128]

Just before Meeting Gorillas—Mrs. Bradley and Porters

[page 128]

very distinctly against the golden green of the meadow, for the opening was then flooded with sun.

Poor Aluma could not understand our inaction. He put his hand on the gun on my knee and tried to urge me forward. He could see no mortal reason in streaking through the forest for hours, stung by nettles and stiff with fatigue, if, after you caught up to your game, you did nothing about it but sit down and look.

Some minutes later he caught my arm excitedly, whispered, "Kubwa" (big one), and pointed to the woods in front of us, not to the left where the others had gone. We could see nothing but the waving of some faraway bushes, which might have meant antelope. Aluma insisted that he had seen a big one, and we were whispering about that, in checkered English and Swahili, when squarely out of the woods at our right a gorilla came in a great hurry and streaked through the bushes into the meadows beyond, like a half-submerged submarine plowing through green waves.

All I could see was the great black hulk and swinging arms—the sunlight on the back was deceiving, but it seemed to me gray. The females are black while the males have the distinguishing silvery hairs. After the gorilla had disappeared, we heard several barks, like exaggerated dog barks.

At any rate, there were five gorillas, six, if Aluma could be trusted, feeding there just out of sight, and we kept looking from the forest to our watches and making futile estimates of the porter's speed, and the time of Mr. Akeley's arrival.

We sent Aluma back to the others and Herbert and

I proceeded to lunch somewhat sketchily upon chocolate and cold beans. Once more we saw the black dots cross the opening, far ahead this time, and we thought of preserving our bean can for Mr. Heinz as the only variety eaten in sight of a wild gorilla.

After an hour, Herbert went back to start another runner on the way. As I sat scooping up the beans I saw something begin to materialize on the slopes to the left and got a glimpse of black coming down the hill before it disappeared into the welter of bushes below. It might have been one of the former band or it might have been a new one.

At any rate I had a strong conviction that the place was boiling with gorillas, and I took a cautious look to see if my Springfield was on Ready and not on Safe. And then I heard a noise at my extreme right.

It was a very loud noise. It sounded just on the other side of the thicket.

The only way I could see at that angle was to rise and look over the top. . . . The crackling came again. . . . Something *was* coming. . . . I hastily recalled that little bit about gorilla hunting which told that the gorilla invariably paused in its advance to beat its breast and roar, and the hunter was advised to hold his fire until a desirable nearness had been obtained.

The question at the moment was what constituted a desirable nearness. The stealthy, twig-snapping sounded quite near enough, although there had been no preamble of warning roars. And the reach of the gorilla's arms was considerably longer than my own.

I had a sudden vision of that arm of the big bull of Karisimbi . . . and I remembered another little bit about the derisive way the gorilla was said to have of snatching away your weapon and snapping the barrel just for exhibition purposes, before dealing its death-blows.

I did all this remembering in about an instant of time, while I was rising out of the thicket, my gun pointed. In that instant I got the keenest thrills of gorilla hunting that I experienced.

I looked out, expecting to see something big and black. . . . A sort of partridge whirred up. . . .

I sat down and finished the beans in my reaction, and when Herbert came back there were no beans but merely an episode.

At two-thirty Mr. Akeley arrived with Miss Miller and the camera, accompanied by the guides who had been sent back by their sultan in great displeasure at their having abandoned the white man before he was through with his work. Good old Burunga! Mr. Akeley certainly endowed him for life with backsheesh. We wondered how much of it the guides themselves ever got.

We had heard the gorillas an hour before the guides arrived and I had seen that lone one within the hour, so we knew we were not far behind them, and we started out with high hopes. But we reckoned without the guides. They did not start along the trail. Instead, as they knew the country, they skirted the high edges and half an hour later cut down into the woods at a place where they hoped the gorillas would be—but the gorillas had been traveling. The guides struck the trail in the meadow and, instead of going on, followed it back to the very place where we had first seen them.

It is very difficult to determine which way a gorilla is going, for the dense vegetation shows no marks of hands and feet, and you are chiefly guided by the swath that his great body makes in the bushes and the broken stalks of the plants he has been feeding upon.

It was then three-thirty and we had lost an hour. Aluma, who had been yearning to distinguish himself by guiding, was nearly weeping with rage, and I sympathized with his more direct methods. Now we followed the trail forward and wandered on and on, up and down, circling great clumps of trees, and winding in and out of dense shrubbery.

There were fresh marks on the trail, also other, and older, ones. We passed many gorilla beds, some old, some recent. The gorilla makes a very simple sort of nest for sleeping—a scratched-together, temporary affair, which he is under no pains to keep clean. There was no skill shown in the construction of the nest; very often, but not necessarily, it took advantage of some hollow at the base of a tree.

The trail we were following became more and more strenuous. The gorillas showed supreme disregard for impediments; they had an underground way of sliding through tangled tree roots or tunneling under branches that was distinctly hard on our hands and knees, but the excitement of the chase sustained us, and we stooped and slid and tunneled as fast as we could.

As hope waned, our criticism of the gorilla's wandering and inconclusive habits deepened bitterly. He was unstable, erratic, and capricious! If he wanted to get anywhere why on earth didn't he go there, instead of rambling and whipsawing all over a mountainside!

Along the way we kept seeing single trails branching off as one after another of the band abandoned the main trail, but the guides were not to be diverted, not even when we passed a muddy way, where the print of recent knuckles and feet were clearly shown. It was the first print of a gorilla I had seen since the day Herbert and I had followed the mud chute up the Mission forests. Here in the highlands there were no other paths where the earth showed through the greenery to reveal a mark.

We were rather inclined to this clearly defined spoor, but the guides held to the main thoroughfare and by and by I began to understand the reason why, for along the way they visited their native snares—a bent pole with a loop for unwary antelope. I was nearly caught in one of them myself, but otherwise there were none of them sprung.

The fact that the natives came up to these forests for wood and for these snares armed only with their spears showed that they were not in great fear of being attacked by the gorillas. I heard of only one case of a gorilla attacking a native here. That was of a woodcutter who was working obliviously when a gorilla sprang out and bit him furiously in the leg, then flung away again, without further attack. The man, fortunately, was too terrified to make any attempt to defend himself. He afterwards recovered, and the incident certainly did not prevent his fellows from coming as freely as before into the woods—any more than a seizure by crocodiles keeps the blacks out of the rivers.

The sky was growing darker, with swift gusts of rain. The hope of a picture was gone and the guides were now hurrying along a path that clearly led to the camp.

It was very thoughtful of the gorillas. We could see no signs of their travel on that way, but the guides insisted that they had followed the trail persistently, only just in front of the camp the gorillas had left the trail and gone down a precipitous cliff across to the opposite mountain.

Perhaps they had. At any rate the picture hunt was over. It had been a vain chase for Mr. Akeley and Miss Miller who had not had the sustaining excitement of seeing the band in the first place. But if the gorillas were aloof, the exercise, Martha assured us gayly, had been unlimited. Martha had a pluck and cheer that no climb or climate daunted.

The loss of the picture chance was not as disappointing as it would have been if Mr. Akeley had not already realized the dream with which he had brought his camera up to the mountains, for he had secured some hundreds of feet of motion picture film of wild gorillas—the first motion pictures of savage gorillas ever taken. He wanted to supplement this with more gorilla film and on another day, with Mr. Bradley, he came on a band, with some big males, that he was able to photograph.

To Herbert and myself it had been a wonderful day.

The Peak of Mt. Mikeno, 14,600 Feet

[page 130]

Porters' Huts at the Gorilla Camp

[page 130]

A Glade in the Bamboos
[page 131]

Native Cattle in the Foothills of Mikeno
[page 131]

I had seen six gorillas, one of them, at least, and probably two, the demon male, and five gorillas had certainly seen us. And we had not been attacked on sight. Not one had beat his breast or roared or tried to ambush us! That is our evidence, as far as it goes. When wounded or cornered the gorilla would be as terrible an antagonist as a giant of such strength and intelligence would naturally be, but we had no reason in the world to believe that the gorilla hunts man, or attacks him unprovoked, or carries off women as in the good old story book tales.

Here in the mountains the natives were constantly going up into the forests frequented by the gorillas and yet I heard only one story of attack—that of the woodcutter in the bamboos—and that might have been some youthful gorilla's idea of a prank, or the ape might have been taken by surprise and believed himself attacked. We never heard of any of the native shambas being raided by the gorillas, for here the gorilla food—the wild carrot and parsley and fresh succulent green growth of the high lands—was extraordinarily plentiful, and, below the upper forest, were the bamboos whose fresh green tips were always in season. The gorilla is a strict vegetarian like the elephant and buffalo—three of the four most dangerous animals of Africa. It behooves one to walk softly with vegetarians!

Altogether Mr. Akeley's belief in the essential character of the gorilla was justified. He was simply the big monkey, the man ape, powerful beyond all words, dangerous when attacked, but not a bit the hellish demon or the malignant arch fiend!

There is no excuse for keeping the gorilla on the game lists. He is too valuable and too rare to be exterminated. He ought to have his own preserves and official protection on his mountain heights and if he doesn't have them, and that soon, he will go the way that so many great beasts have gone—the way that all are going fast now in Africa. We estimated that not more than seventy-five or a hundred of the gorillas exist in those mountains. Though our licenses gave us ten gorillas, we killed only the five necessary for the Museum group.

It is extraordinary that the gorillas are not more numerous in this area for, until the last years brought the hunters I have mentioned, the great apes were entirely unmolested. They breed undisturbed, but evidently they do not breed very fast—or else some unknown cause of mortality keeps the numbers down. That can hardly be credited, for there was not a hint of disease in any gorilla dissected; nor was a solitary parasite found in any gorilla, nor yet on a gorilla. Their hair was as soft and pure and free from insects as a freshly tubbed pet kitten. It is all the more astonishing when you think of the literal millions of ticks that are on lions and elephants and buffaloes and rhinos, and apparently every other jungle creature. Famine seems as much out of the question as disease, for if the food of these upland meadows failed, the gorilla had only to descend to the bamboo forests below where he would find plenty of fresh shoots. The fact that the gorilla had never gone down to the native shambas showed that food had not been a problem. Lacking any other factors in the situation, it must simply be that the rate of increase is extremely slow. As far as is known the births are always single. The mother that Mr. Akeley photographed had two little ones with her of much the same size and from that one might conjecture them to be twins, but it would be sheer conjecture—no record of gorilla twins is known.

Their longevity is said to be greater than that of man. Mr. Barns thought that his male gorilla had lived a hundred years, but he was frankly voicing a possibility. Mr. Akeley was inclined to think that the age limit was more nearly that of man. This question would be one of the interesting things that could be determined by an experiment station of naturalists in gorilla land.

The organization seems to be the band rather than the family; our experience that noon and, later, the experience of Mr. Akeley and Mr. Bradley with a large band on the slopes of Mikeno, showed that the group might consist of two or more males with a larger proportion of females. The question arises whether those bands consisted of two or more respectable monogamous couples and their marriageable daughters—maiden gorillas yet unculled by roving gallants—or whether it consisted of a couple of gorilla gentlemen and their respective harems or of unassorted and liberally inclined ladies and gentlemen. . . . We can only offer the situation, not the solution.

The gorilla sleeps on beds and not in trees as do the chimpanzees in bamboos, and whether they use the trees at all, except to climb up a slanting trunk to a crotch, is a question which our evidence would have to answer in the negative. Mr. Akeley saw gorillas in the crotches of trees and so did the Père Van Hoef, but that simply means that the great apes had climbed up to a comfortable perch as a youngster climbs. It would be impossible for them to swing through the trees like their monkey kin, for the enormous weight of the beasts makes that legend untenable. Very few branches could support from three to four hundred pounds!

The tremendous strength of the gorilla is a mystery. Where does he get it and why does he have it? Not in a necessary circumstance of his life does he use it now. Those great shoulders and bulging arm muscles that could crush a lion have no more arduous work than breaking off wild parsley and scratching together branches for a nest.

The morning after our experience with the band, Mr. Akeley and Mr. Bradley planned to go back after pictures and ran onto a band on the slopes of Mikeno while Martha Miller and I went down the mountain to rejoin Alice and Priscilla Hall. We made the descent in five and a half hours, with a constantly increasing wonder at ourselves that we ever made the ascent at all.

Out from the Mission an excited little girl with flying curls came racing to meet us and Priscilla had cooling lime juice ready for us, then a midday meal that was the height of luxury—roast chicken and green corn from Burunga, and Cape gooseberries, strawberries and bananas from the Fathers' gardens; then came hot baths and clean clothes, and another dinner that night at the hospitable Fathers'. Lastly we slept in our white-washed room with thatched roofs and mud floors with a sense of comfortable homecoming.

We had attained our objective—that high triangle which we had believed covered with bamboos and which proved such an enchanted forest—and we should always carry with us a picture of the wonderland of those hidden heights, the great, cloud-wrapped forests and their giant denizens.

It had been worth it all; the heart-breaking climb, the cold, the discomforts were merely incidents, a price one paid gladly—especially in retrospect!—for the rare experience of seeing gorillas in their savage solitudes.

CHAPTER X

THE PYGMIES COME TO CAMP

THANKSGIVING WITH THE WHITE FATHERS

THE Batwa were coming. The Pygmies, the fugitive little forest dwellers, were coming to our camp.

A hasty message from the White Fathers brought the word, and I was so afraid that they would change their fugitive little forest minds that I seized Alice, our star attraction with the natives, and with Martha and Priscilla hurried to meet them.

Ever since we had reached Lulenga, before we went up the mountains after gorillas, we had been trying through the kindly offices of the White Fathers to get the Batwa to come in. They are the forest dwarfs, the descendants, in all probability, of those little bushmen who succeeded the kin of the Neanderthal giants, and who, a hundred thousand years ago, while Europe lay under glacial ice, were covering the rocks of Africa with their delicate paintings and drawings.

They were nomads, shy, suspicious, often at war with the blacks. Later, in Uganda, we found that the Pygmy raids from the forest had wiped out the population of the southern shores of Lake Bunyoni, and though the British had conducted a punitive expedition against the Batwa, the natives did not settle there again. But here the relation between the dwarfs and the Banyaruanda, the inhabitants of this region, was friendly, even to intermarriage, and the White Fathers had induced the chief to send into the forest for the Batwa, with a promise of salt from us for each one that would come in.

Mr. Akeley and Mr. Bradley were still on Mikeno, planning to reach camp by noon, for it was Thanksgiving Day, and as we hurried along towards the Mission we felt an anxious pang for fear the retiring Batwa would not stay to meet the men and be recorded in the motion pictures.

We might have spared that pang. Ahead of us we saw coming, not the shy two or three, advancing cautiously from bush to bush as our fancy had pictured, but an actual procession pouring along like a pilgrimage, clapping and singing and followed by a mob of curious natives.

Martha and I exchanged bewilderment. Were these our shrinking Batwa? . . . It could not be.

It was the Batwa. Men, women, and children, the tribe of them was making holiday to see the strange white women and the child with the long, fair curls—and to receive the promised salt. Short, stocky little dwarfs they were, with round heads which had the effect of being driven too deeply between their shoulders, clad simply in leather aprons, the women supplementing the aprons with copper wire anklets and a head wreath of flowers, the men carrying bows and arrows. The children were unencumbered.

Not that that was any distinction. No little African has to be careful of his clothes.

They poured into the cleared space before us and ringed Alice in, chattering their monkeylike clicks and grunts, and then they danced. They danced like mad for three hours. We tried to get them to wait till the Bwanas had come down the mountain, but no, the spirit was on them, and they clapped and sang and jazzed in oblivious excitement, while all the natives in the vicinity gathered and clapped with them, and in that pandemonium we finished breakfast and the preparation for Thanksgiving dinner.

When the men finally appeared, our Batwa had exhausted their first fine frenzy. For the sake of the salt, of which they are inordinately fond, they repeated the dance, but it was a much less spirited affair.

These were the Batwa of the forests of the volcanoes. The name Batwa — Mutwa in the singular — means simply "Little Men." We had heard of three Batwa peoples in the Eastern Congo, the tribes in the Ruwenzori forests, a tribe upon the Island of Kwijwi, the largest island on Lake Kivu, and these of the Eastern forests. The Kwijwi and Ruwenzori Batwa are identical with the true Pygmies, from which all the Batwa are generally held to be descended; these of the M'fumbiro region present the dwarfish characteristics—the rounded head, the big eyes, the broad-rooted nose—but intermarriage with the friendly Banyaruanda, the neighboring tribe, has undoubtedly modified the stock, especially in the matter of height.

The Duke of Mecklenburg reported that he found these Batwa much taller than he expected, some were four feet seven, some even four feet eleven. We found some men about that height—and in one case we found that there had been intermarriage in that family with the family of a negro chief—but other men were barely over four feet high, and the women, who were the ugliest creatures that I saw in Africa, were smaller, squat, and thick. We did not try to measure them for fear of intimidating them, but we had Alice stand by them, and measured the difference. Alice's height was three feet eight inches and there were women with children on their backs who were no taller and two much shorter. Their color was black, warm-toned, and many of the men seemed hairier than the negroes.

Their origin and history are part of the mystery of Africa. They say of themselves that they were "always here," the oldest inhabitants of the region. It has been considered a sign of the comparatively recent origin of the Bantu tribes that they live upon imported grains, while the forest dwarfs subsist on native fare. When there is a famine, as there was after the Great War, the Negro tribes die in millions, but the Batwa with their spears and bows and poisoned arrows rove the jungles and find food. They eat birds, rodents, and small game; ants and caterpillars are not disdained; and nuts and berries and the roots of ordinary trees are helpful. Dr. Van den Bergh, a White Father of Uganda who journeyed to the Congo forests west of Albert Nyanza, stated that there they ate only the meat which the forest supplied and which was killed by themselves, or at the killing of which they had assisted, or at least had shared in the tracking, but this is not true of the Batwa of South Africa. There, General Smuts told me, the little men live often upon carrion, and one would lie out in the bush for hours, his unwinking eyes fixed upon the vultures high overhead, invisible to other eyes less keen; and when the vultures sighted a dying animal or a lion's kill, and began sailing towards it, the bushman would jump up and run to it, and after the lions and jackals had feasted would creep in and pick the bones.

A man known to General Smuts had put out a carrion bait and lain in wait, and so had succeeded in capturing two Pygmy children. He kept them six months, but in spite of all he could do, they pined and sickened and, for fear that they would die, he let them go.

The Batwa were all nomads, ranging their forests, building temporary huts of twigs and branches, and seldom staying in them longer than two weeks. They cultivated no fields, no shambas; they traded a little with the natives, usually for tobacco. Like all the natives of the region the men carried black clay pipes with wooden stems; these pipes were very finely made with something of the style and finish of the old Dutch ones. They obtained most of these pipes from a native of the vicinity who made them, but the Batwa East of Kivu did make unglazed pottery of their own, either black or red, of which we had obtained some samples through the administrators we had met.

Far from being excruciatingly shy they became so interested in us and our domestic details that they were underfoot like inquisitive kittens, and Herbert hastened to present them with the promised salt. We gave each one a good tablespoonful and a bowlful to the chief. The chief showed a strong sense of honesty, for he

Photograph by Père Provoost

BATWA CHIEF AND WIFE

[page 140]

DANCE OF THE BATWA

[page 140]

Crossing the Lava Plains

[page 150]

A Night on the Summit

[page 150]

brought back three women who had not danced for us and made them give up their salt. We then presented salt to the native chief who had induced them to come in, and shook hands ceremoniously with all the chiefs we were sure of, and several who seemed to be sure of themselves, and wished them all "Kwaheri." Then, rest in camp being merely change of occupation, the porters were lined up and paid off, with all the promised backsheesh. And then we had a real dance.

The porters started down the slope, turned suddenly, with a shout of song that rose and fell like hammer-strokes, and came up the hill to us, their four headmen in the lead. I never saw men dance as those men danced. All their joy at release, their triumph of danger overcome, their exultation at fortunes made and the wives to buy, was in that leaping, stamping orgy. Four times they surged up to us and four times receded, a marvel of wild and savage motion.

Thanksgiving dinner was on the porch of the mud house. It was a day like June, with only a sprinkle of rain—Thanksgiving with roses blooming and volcanoes glowing. Thanksgiving with chicken instead of turkey, and strawberries and Cape gooseberries from the Mission instead of cranberries, with green corn from Burunga's shambas, and plum pudding instead of pumpkin pie—plum pudding with a sauce that does not exist now within the three-mile limit, creamy and buttery and fragrant!

We had two dinners, the family affair at two o'clock, and another at seven to which we invited the White Fathers. Think of saying tranquilly to your cook,

"Two dinners to-day—the first for six people, and the second for ten, and six courses to each dinner!"

It did seem to us a fairly full day for the kitchen, but we did not know our cook.

A six-course dinner—soup, chicken, potatoes, asparagus and corn, salad, pudding, cheese, crackers, coffee, nuts, and candy . . . and four guests. . . . Only six courses. Well, he was a Congo cook. He had served the *chef de poste* at Kissenyi, and he would save us from ourselves and our shame if it could be done.

Heaven knows how he gathered the oddments of ham and flour and lard and unreliable eggs—if I had known his intentions I would have given him the keys and let him do his best and not his worst—but out of his clandestine this and that and what-not, he supplemented those six courses with six others of his own and served a meal that satisfied his ideas of state.

He gave us croquettes of the beef of yesteryear, and vegetables matriculated from the soup; he gave us pie of cooking fat and discarded citrons with a meringe of hoarded eggs! He alternated his courses with the courses of our selection, but not in order, so that the green corn came after the dessert, but it all came, course after course, each with the complete change of plates and knives and forks, and a harrowing wait for the washing of those plates and knives and forks.

Night deepened, stars brightened, volcanoes glowed redder and redder. Candles spluttered and burned out and were replenished. Conversation, our French conversation, spluttered, too, in those awful waits, then revived in gales of laughter.

And still the boys came, serving and serving. . . .

And one of them served, in honor of the occasion, in the union suit that Mr. Akeley had given him on the gorilla mountains to supplement his shirt and khaki shorts. Then he had worn it invisibly, as custom prescribes for union suits, but in these warmer regions Leo did not need so much clothing and discarded the shirt and trousers for the newest thing. Anything quainter than that sketchily buttoned union suit passing the potatoes I have never beheld. Never, when buying the garment in New York, had Mr. Akeley contemplated its appearance behind his dining-room chair!

The other boy, Mablanga, had gone to the other extreme. He had trousers, a coat but no shirt, and he pinned his coat scrupulously across and wound about his neck a woolen muffler.

Merrick, the wash boy, aided the festivities prettily attired in pink pajamas donated by Miss Miller.

And then Aluma, an unpopular young tent boy, the one with us during the gorilla hunt who shirked so badly that he was in disfavor even with our easy-going lot, suddenly appeared on the edge of the outer darkness and in a horridly exaggerated croak from mountain colds, like that of a dying frog, insisted upon enumerating the sorrows of his life to our guests who spoke his language.

At the conclusion of the whining recital they waved him majestically away and all the translation their politeness gave us was that he was dissatisfied with the climate and the gorillas and that the other boys made war on him.

He deserved war and we sent him back to Kissenyi

next day. Discharge is simple in the Congo. You reckon the time it will take the boy to walk back to the place you brought him from and give him thirty cents for each week of the way. If a boat or train is to be had, you give him the native fare.

Our Thanksgiving evening concluded finally with games and gayety—and then from midnight to morning we devoted ourselves to packing and to the arrangements for another expedition on which we planned to start the next day.

CHAPTER XI

A NIGHT IN A CRATER

The Ascension of Mount Nyamlagira, Africa's Most Active Volcano

For weeks there had been the glare of volcanoes on our horizon. We were in the most active volcanic region in Africa, the scene of the most recent upheavals. Here, north of Kivu, the great Central African Rift Valley, which stretches between the mountain walls from the southern end of Lake Tanganyika over Lakes Kivu and Albert Edward to Albert Nyanza, had been dammed at some not distant—geologically speaking—date by the bursting up of the M'fumbiro group of volcanoes, and the waters of Kivu were thus cut off from the Nile basin, and raised to the remarkable height of nearly five thousand feet.

The eight great volcanoes fall into three groups: Sabinio, Mgahinga, and Muhavura stretch east, one behind the other; Mikeno, Karisimbi, and Visoke, our old gorilla friends, form a central triangle; and, on the western side of the valley rise the separate slopes of Chaninagongo and Nyamlagira, the only ones that are now active. All the others have been so long extinct that the natives have no knowledge of them as fire mountains except Muhavura. There the natives refer to an eruptive flue upon the mountainside as "Kabiranjuma" (the Last Boiler).

From the lava flooded floor of the Rift Valley, south and east of Nyamlagira, rise innumerable little cones, many of which have been active or come into being within very recent years, some as late as 1912, quiet now, their rocky sides beginning to green with lichen. Chaninagongo and Nyamlagira were the only guardians of the restless fire and over their flattened crater tops the clouds of steam hung rosy tinted.

Chaninagongo smoked but slightly, but from Nyamlagira went up a cloud of steam by day, and by night a crimson blaze seen a hundred miles away. It was the most active volcano in Africa, probably in the known world, and no one had ever made the ascent since the last outbreak of fire.

In 1904 it had begun to be active; in 1905 Wollaston reported a jet of steam escaping from the southern slopes; in 1907 Kirschstein of Mecklenburg's expedition witnessed eruptions and made the ascent to the crater. For some time then it smoked heavily, as far as we could learn; Monsieur Van de Ghiniste, the Belgian Commissioner at Ruchuru, got up once to the summit but could see nothing on account of the clouds of smoke and fumes. Then fire broke out and no one had been up since that last eruption.

Its ascent was now considered impossible on account of the fumes, but from the Mikeno camp Mr. Akeley had noted that the wind blew steadily southwest, so the other side seemed safe. There was difficulty in approaching that side as the mountain was encircled by a great lava field, an outpouring from a tiny cone in the valley floor.

Yubile was its native name, but after the eruption they called it "Chamunaniyo" (they shall not pass).

That lava field was a mass of broken, porous rocks—churned-about blocks like dynamited pyramids. Plants were finding rootage, and the low waving branches of green gave it the air of a field, but it was a desperate thing to scramble over, and practically impassable for porters with loads.

But Père Van Hoef knew that the natives knew a way across to the opposite forests and when we decided to ascend the volcano, he got us guides and porters, though he had his doubts about the porters making any ascent. The day after Thanksgiving Day, Mr. Akeley, Miss Miller, the two Bradleys, and about forty luckless natives wound down through the banana groves, past the huts, and started to pick their way over the lava field.

It was the world's worst going. The porters made grass sandals for their feet, and clung to stout sticks. They began to yearn volubly for home.

In the Congo your worst fears are never realized. Something that you didn't fear happens instead. We had expected a beating midday sun upon that lava plain—we got a freezing storm of sleet and hail, and the hail was literally the size of hen's eggs—Congo eggs. The Bradley raincoats were on the head of the porter somewhere out of reach—the porter with your personal belongings is always as elusive as Peter Pan, while the one with dried beans is always close at hand—so we had reached a shivering saturation point by the time we found shelter.

We made camp in a glade on the lower slopes of Nyamlagira, and then for two days those guides led us round and round that mountain—anywhere but up.

The sides were cleft with fissures and ravines, so that we had plenty of climbing to do, and they must have cherished the fond hope that if they exercised us sufficiently we would call it an excursion and go home. They had a violent aversion to approaching the crater, and an equally poignant sentiment against walking up the mountainside on an elephant track with our luggage on their heads.

Our second camp was in the jungle, in forests utterly unlike the gorilla forests of Mikeno and Karisimbi. There were no evidences of gorillas here, and none of their food. There were elephants, and we ran into them on the march. We heard them quite near. The green was too dense to see. We waited; there was a profound stillness, the snappings and rumblings ceased and the cessation told that the elephants had seen us and were waiting, too.

Then some branches cracked. The porters pitched their loads—one of the loads being Mr. Akeley's precious camera—and fled. Mr. Akeley shot at random, and the result was what he hoped for: the elephants crashed off. I got a glimpse of a gray back that looked as high as a barn. We would have liked an elephant hunt, but we could not take time from the ascent.

The third day the porters decided to stop. They decided it every hour. At last Mr. Akeley got the kilangozi, the headman, far enough in advance to be out of ear shot and drove him up, while we shepherded the rest, like reluctant chamois. One or two bolted, but the others struggled up the elephant trail, through the jungle, on through barren heights of sparse grass till we found a camping place about half an hour's climb from the summit. We weighted the tents down with lava rock and the porters made huts of grass.

Mr. Akeley went up ahead to the crater, then went up again with us about four-thirty. Martha and I went hand in hand to share whatever distinction there was in being the first women to put our heads over Nyamlagira's fire.

We stood on the brink of a great crater about eight miles in circumference. We looked down into a colossal chasm blown out by three separate eruptions—table lands and bastion walls, cinder slopes and spouting steam and billowing sulphur clouds—a wild, demoniac place that was as beautiful and mad as a volcano ought to be.

The fire came from the third crater, the only active one. We could see the glow and we made our way for three quarters of an hour about the summit, over jagged lava, until we could look through a break in the inner walls into the pit of boiling lava from which came the explosions that thundered up like the roaring of surf. It was a gorgeous sight—the fire of the very heart of the world at our feet and all about ethereal distance and great mountains melting into space.

East of us, across the wide valley, were Mikeno and Karisimbi, purple peaks with clouds streaming like banners from their pinnacles, and Visoke, the third mountain of the gorilla triangle, just visible on the left, and yet farther to the left Sabinio, Mgahinga, Muhavura, so directly in line that they seemed one mountain. Karisimbi was the highest of all—14,638 feet—and a faint sprinkling of snow was powdering its peak.

Northwards stretched the valley, a shadowy reach of forest and lava plains, guarded east and west by mountain ranges; south were the slopes of Chaninagongo, and across from us, like an azure cloud against the crater's black edge, lay distant Kivu.

We looked too long. The sun set, night swooped down like a bat, and up the mountainsides rolled enveloping white cotton clouds of fog. We tried to save time by taking an oblique way down to camp and for three hours we floundered in fog and blackness and rocks and ravines and brush.

It rained. It poured. We huddled under a bush until the fury of the storm was spent, feeling we were taking refuge in a watercourse, so tremendous a fall of water washed down that mountain's sides. Then we kept moving to keep warm. Ultimately an answering "hallo" from camp brought us down to our welcoming boys who were fairly pallid with fear that the devils of the volcano had us!

The night ended in hot chocolate and dry dressing gowns. We had brought but one change of clothes on this volcanic expedition and both my suits had now been caught out in storms and hung dripping from a line in my tent. It seemed to me the next morning that they were wetter than when I had hung them up for the mountain dew had condensed like additional rain. I hesitated between them, but there was small choice. Anyone who knows the delight of drawing on a wet bathing suit on a chilly sunless morning knows the sensation. Yet we never had a cold nor felt the least ill effect from this dampness.

We went up to the crater next morning, planning to spend the night at the point on the summit where we had stood the night before. We sent some porters to the left to carry firewood and a chop box and some blankets and bags to that place and a piece of canvas for protection, and we ourselves, with other porters carrying our cameras, started on an exploration trip to the right. We made our way slowly along the rim, at first over upthrust rocks which made the going frightfully difficult; then we reached a place quite smooth with cinders, and marched until a dip in the rim enabled us to climb down to the level of an older crater floor, billowing white with clouds from yellow and green sulphur beds.

Three distinct eruptions had blown the interior of the crater into three giant abysses separated by bastion-like walls of rock, stratified and colored like a mammoth layer cake. Almost in the center rose a citadel rock—the "Castle," we called it, amber in the sun. The terrace on which we gained admission to the interior was an old crater floor, blown up by the more recent eruptions; it was ominously hollow sounding and we started across it with commendable prudence, clinging to a rope and testing every step. But after we found the crust supporting and the sulphur beds harmless, we grew fairly blasé.

That terrace ended abruptly in a clifflike drop with cinder slopes and flats far below; we skirted the edge until we found a place from which we could clamber down to a little ledge, boulder strewn and warm with issuing steam jets, leading to the left; to the right a long cinder slope stretched down toward the boiling lava; we saw that the cinders ended in another drop. A brief rain overtook us here and the unhappy camera attendants found opportune refuge among the boulders and warm recesses. Then we retraced our steps across the flats to the break in the rim, and continued the circumference of the main crater.

We estimated later that it was about eight miles around the crater. The Chaninagongo crater which has been several times ascended is held to be five miles around. For a time our way was up and down among cinders and boulders; then the rain, and this time a real rain, overtook us, and we found shelter in a huge cave, full of interesting stalactites. There was a warm cave next to it which attracted our camera bearers. As soon as it cleared we completed the journey, now over cinders, now over rough porous lava rock, till late in the afternoon we reached the place on the eastern side from which the night before we had looked down into the blazing lava between the portals of high rock.

That night we camped there, beside a boulder, beneath a canvas stretching from the boulder to the ground, weighted down with rock. The porters descended to the base camp, gladdened with a franc apiece for backsheesh, and after we had watched the distant fire and photographed it, we laid us down to sleep upon lava rocks lightly overlaid with torn up grass.

A NIGHT IN A CRATER

It was a cold night. We had dressed elaborately for it, discarding nothing and adding everything in the line of sweaters and raincoats that we had. It was none too much. That we slept was due to the day's generous exercise rather than any coöperation on the part of the rocky beds and the weather.

Sunrise was a gleam of gold behind amethystine peaks, and a sea of cloud rolling through the valleys. During breakfast our porters appeared with water and firewood, faithful to the rendezvous, although from the new faces among them we judged several originals were unwilling to brave the volcano again even for a franc.

The cook came up with them from the base camp, out of curiosity to see the devil's fire, but he was too late to do anything for us but clean up; we had to be our own cooks on this camping party, and we learned what it meant to get any sort of meal on a fire between three stones. I knew then why the cook's eyes always looked red and inflamed. Our firewood on the summit was particularly trying for they had only flimsy stuff to bring up to us, and we supplemented that with brittle twigs gathered on the heights. The vegetation on the top was sparse. There were a few canelike stalks of stunted trees and a little brushwood down the slopes, but above that were only coarse grasses and white everlastings, and tiny, ground-gripping creepers with small bright blossoms.

We went around the rim again to the spot from which we could descend into the crater, and marched out across the flats for half an hour, feeling well acquainted now with sulphur beds and not blanching when an unwary foot broke through the crust—the fact that man and woman are two-legged beings saved some of us from total eclipse—and reached the edge of the terrace and skirted it to a point from which we could make a further descent to a narrow ledge below. To the left this ledge clung against the precipicelike walls; to the right were cinder slopes stretching dizzily down to end abruptly; and beyond the cinder slopes jutted a flat-topped cliff.

On that ledge against the crater wall we stretched our shelter cloth, weighted the lower edge of it with tied-on rocks and dropped them over the brink. There was steam pouring from the cracks near us, and sifting up from the gravel whenever our heels scratched it. We had no fear of being cold that night.

Here we made ourselves at home. The porters clutched their francs and took what they clearly thought the last look at the mad ones, and escaped. Mr. Akeley proceeded to climb down the cinder slope to the right and set his camera on the edge of nothing. We could look down now into the crater pit of boiling lava, a great mass crusted with cooler, darker lava, patterned with gleaming cracks of pure gold that shifted and changed as the stuff seethed and boiled and broke into fountains of fire or rolled into molten rivers of hissing flame.

As night came on, the cloud above that cauldron became a glow of rose, vivid, unearthly, a rose of hell—a drifting, billowing radiance that made an Inferno of those rocky walls, throwing into weird relief every jutting ledge and rock, filling with mysterious shadow the

LAVA IN ERUPTION IN NYAMLAGIRA'S CRATER

[page 154]

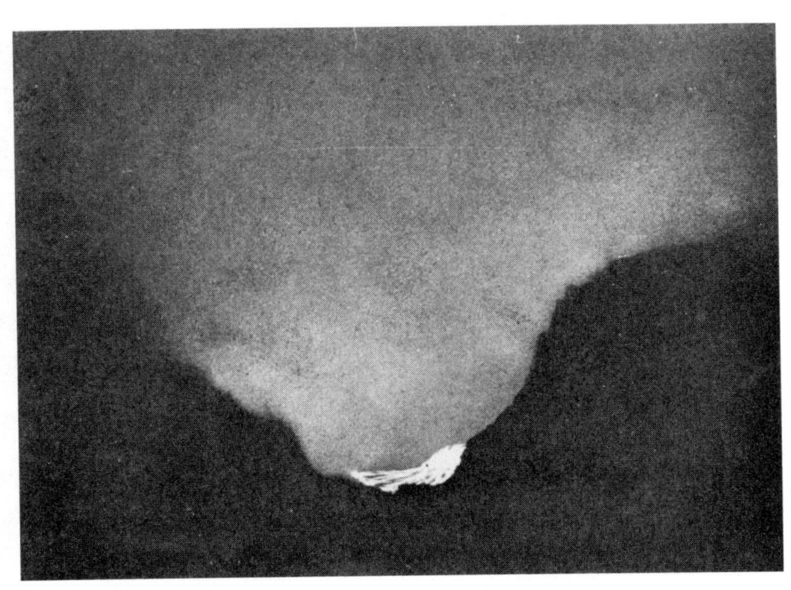

THE FIRE POT OF NYAMLAGIRA—TAKEN AT NIGHT BY ITS OWN LIGHT

[page 154]

Sunny Jim

[page 158]

Our Bicycle Boys

[page 158]

deep recesses and dark distance. High above the crater that rose-red cloud streamed out against a sky its fire made black. Half the night we sat out upon that ledge, the glory of that spectacle filling our eyes, the thunder of it in our ears. . . . It was for us alone; we had been the first to look upon it. And for night after night, year after year, that glory was blazing and thundering away, unseen, undreamed. . . .

It was an unforgettable night. It was Rome burning. It was Valhalla ablaze.

The next day we crossed the cinder slopes and went to the edge of the lower terrace, where we waited hours for the fogs to lift from the inner chasm so we could use the cameras; then at noon Martha Miller and I started back to camp, our green bags on the heads of two porters and a boy, Leo, for guide. After our experience that first night Martha and I had a strong feeling that it was best to go around the crater to a spot just over camp, then descend in a straight line, but the porters naturally hated the hard going over the stabbing rock edges of the rim, and struck down on a diagonal line. We thought that we could trust to their sense of direction; they were natives, they ought to be endowed with native instinct for direction; moreover Leo, the boy, had been coming up every day. Lightheartedly we followed them and emerged from a stretch of brush to find one porter on one hillock descending to the left, and another on another hillock streaking it for the right, and the boy Leo, a lost mariner on unchartered seas, shrieking questions at them both.

And then, naturally, it rained. Bitterly we ceased

to trust to their native instincts, and relying on our own we herded them ahead of us and ultimately gained the camp. Their deviations had cost us over an hour of hard going. We told them what we thought of them in all our English and Swahili, and then they presented themselves sheepishly smiling, for a present for their hard work! They didn't get that present.

At four the men came down, having tried all day for pictures and been thwarted by fog and clouds; that night was the coldest in the history of the world. Our provisions were running low; we could have hung on, ourselves, with what we had, and we had splendid mountain streams of water, but the beans on which the porters subsisted were only enough to get us back to the Mission, so we started down the mountain the next day, our porters jubilant at leaving the Sheitani, the fire devils, behind and eager to boast their experience in their villages. Alone in Africa, they and we had looked into the pit of Nyamlagira's fire.

CHAPTER XII

THE THREE-STONE KITCHEN

HOUSEKEEPING, on an African safari, was a never-ending occupation. Our experiences had begun on the Lualaba River when we were precipitated violently into the need of a cook, and our boy Mablanga assumed the rôle until we found a regular one. Mablanga cooked well but he did not claim to be an "impishi" (a cook). That was his mistake—and ours. If he had, we never would have allowed him to resign, but perhaps it was as well, for there were other cooks and there was no such useful general boy as Mablanga.

At our first camp on the Upper Congo, Kabalo, we had started getting a cook, and our first candidate was Jim. Jim might be described as husky. He had the robustness that would have made him king of Kentucky roustabouts, and a smile that would have won him a place in the Big Time. Jim was a cook. He said so, himself. We gave him a day to prove it, challenging him to breadmaking, and after we had seen the bread we gave him his choice of staying on as a tent boy or separating himself from any further obligation of service. Jim stayed.

He developed, it may be said, not as a cook, but as a sort of over-boy. He was as strong as an ox, but he seemed to be preserving his strength for some crisis which never came—he produced others, who did whatever work we were calling for. Where he got them was a mystery, but he never failed. Of course, on the march, where there were porters available, life was simplified for him; and pails were filled and pegs tightened and bags brought in by easily available hands; but no matter where we were, forests, mountains, plains, Jim never failed of his supply of help. He had always a "moké" (a boy) or two, some little grinning pickaninny, squatting about with him, running for water, dubbining the shoes, and hanging the wash on the line. Of course there were a few things that Jim could not escape doing in person, but to do him justice they were very few.

I own to a weakness for Jim. I liked to ascribe the whipping he got from the soldiers at Kissenyi to his belligerent loyalty. When I had the fever Jim was sent to Nyunde Mission for vegetables and oranges for me, and a boy from the *chef de poste* went for things for him. On their return there was some dispute over the oranges and when the other boy tried to take fifty for his master Jim resented the attempt especially as he had carried the oranges—and hit the boy. I expect he hit him hard. The next day the native soldiers came and led Jim to an official whipping and we never saw Jim in trousers again. They were new trousers, too. He had had fifteen francs, three weeks' salary, advanced with which to buy them, for the garb in which he came to us was chiefly patches and a huge safety pin. After the Kissenyi episode Jim girt himself kilt-wise with a tea towel, and anon, when his shirt was beyond all aid of thread or safety pin, he split another ancient towel, thrust his head through, tucked in the ends, and was arrayed.

But to go back to the cook and Kabalo—our next experiment after Jim's demotion, was Antoine. Antoine did cook. He made bread of sorts and mayonnaise, for which he went secretly and begged vinegar from the administrator's wife, who was yet unknown to me. He was enterprising and furtive and he had an assistant of a villainous cast, and when the Kabalo administrator observed the precious pair, he warned us that they were probably robbers and not worth harboring.

We didn't like Antoine ourselves and his bread was nothing to cling to, so we explained this to him through Mr. Akeley. Mr. Akeley spoke a vehement Swahili while Antoine knew no Swahili and only a pidgin variety of French. Certainly something percolated to Antoine for he was absent from camp that night when we returned from a late dinner, and there was no breakfast for us next morning before our six o'clock train to Tanganyika left, but at train time Antoine turned up, with his assistant and his wife and a bundle—and we weakly gave him three francs for solace and departed.

At Albertville, our next camp, the administrator sent us a cook. Mr. Akeley liked his looks. He said so and he seldom was as sorry for anything else he said. For we kept the cook, and the cook's assistant, who played perpetually on a small musical instrument, the m'bichi or native piano, and we kept too, another tent-and-table boy who presented himself with them.

He was a shrimpish boy in a yellow and black blazer that some one had given him and his name we took to be Sorrow and then amended to Shallow but either was inspired. The cook may have had a name but he did not need it. We found others. A more utterly witless and useless individual never sat under a tree and drank pombe. Unfortunately we were well on our way before we had a chance to discover his proclivities, and after we had disembarked from the Tanganyika boat and landed at Usumbura he developed fever, a real fever, which shielded his disabilities for some time, although the moral flaws soon became apparent as related in the episode of the visiting lady.

On the safari up from Tanganyika we tried not to be too hard on the cook and the cooking. Most of our food then came from the chop boxes and took little preparation, but his own handiwork was nothing short of a calamity.

His bread grew soggier and soggier. Part of this was due to misunderstanding. He wanted "dower" or medicine, meaning native pombe, to make the bread rise, and Mr. Akeley was of the opinion that the flour he had ordered was of the self-rising sort, only needing mixing with water, and so we went about like parrots, saying "pana dower" (no medicine) to the cook and getting the English-and-Swahili-speaking boy to make it clearer, and every day the cook brought us heavier and stranger bits of dough sunk depressedly into some cavernous receptacle.

Mr. Akeley explained that it was the lack of sun over a tent fly. His bread in East Africa had always risen beneath a tent fly. But after we had mixed the stuff ourselves with water and set it to rise beneath a tent fly—and buried it where the boys could not see—we had our suspicion, and I never said "pana dower" from that moment but ordered the native banana beer.

However, if the cook's bread were not enough to convict him, his meat was a criminal offense. We gave him everything—chicken, oxen, elephant. We didn't blame him for the goat; we didn't blame him for the elephant, we didn't blame him for those chickens which chiefs had selected for presents because of their long and distinguished field service, and we didn't blame him wholly for the first ox—but the second was conclusive.

We had reached Kissenyi, then, where we had fresh meat and vegetables in abundance, and on their behalf we interposed. It wasn't humane to allow even a cabbage to be treated as that cook treated it. We remonstrated and he detached himself from all responsibility, withdrawing into a Buddhistic revery about meal time, letting the other boys do the work. He was accompanied in his Nirvana by his assistant and Shallow.

The night after Jim had his falling out with the government, the other boys, in terror of the heavy-handed possibilities of the post, appeared in a body asking for a dismissal which we totally refused. Then it dawned upon us that we had been undiscriminating. We decided to part with the cook and the cook's helper and with Shallow, and in another day told them that they could go. But they would not go. They had changed their minds about Kissenyi and had no notion of leaving us. They regarded themselves as permanent investments, paying no dividends, but untransferable. Mr. Bradley actually forced the money upon them, and thrust them from the place, yelling like banshees. Then they hung about waiting for the twice-a-month launch, and when it appeared it was unable to take them on, so they started back on foot to the melancholy music of the assistant cook's piano. As a matter of fact they had a trip through lovely country, with plenty of money to make it. Our hearts were not wrung for them and their departure had a bracing influence upon the camp.

We received a new impishi from Monsieur Wera, the Kissenyi *chef de poste,* who was discarding him after nine months for a better one. He was fair enough, he said, but a robber, and we must look out for butter and salt.

I believed every disparaging thing said about the donation. A more uninviting face I have seldom seen than the mistrustful defiance of the impishi. However, we installed him over the three stones that made a kitchen, before the tent where the boys slept and wherein they cooked when it rained, and he installed as a helper a protuberant youngster whose figure was ruined by bananas, whom he paid out of his own forty francs a month and food money.

Vigorously he seized on the meats and vegetables and did wonders with them. He gave us dinners, real Belgian dinners with two meats—often the same meat though differently prepared—and then he went to jail.

He went suddenly. It developed that there was a

captain of the launch at Kissenyi and the captain's black boy had a black lady for whom the cook had a fancy. And the cook had just been paid. Knowing his face I can only attribute his success to his pocketbook. Certainly the next morning when he was fined twenty francs for a fight with the black lady's revengful first possessor, a fight in which white man's property was destroyed, the cook had no francs left. We promised the *chef de poste* to send them back as soon as the cook had earned them and so we did, and the cook went on with us and remained with us throughout.

So did his proclivities. Cook's sisters. He found several along the route and one of the first activities at Lulenga Mission was a descent by Herbert and the White Brother upon the kitchen house, to rid it of its giggling, bead and calico encumbrances. Nominally that cook counted as a native Christian. He wore a medal of the Virgin and he went to Mass. Some day he intended to take a wife and one only. Meanwhile he was undecided as to her indentity.

But whatever his romantic or religious excursions, that cook was a cook and a worker; and the masterly way he got into camp on the march, set up his tent, and had his fire going between the three stones inspired a just respect. We accepted him as a permanent institution and we accepted, too, his red mattress which went from camp to camp on the head of a staggering porter. It was the most enormous mattress that I have ever seen, and mysteriously, from day to day, it assumed more and more swollen proportions.

Watching it being heavily hoisted into position upon

the head of a protesting native the morning of the start for Nyamlagira, Mr. Akeley thoughtfully called Kiani.

"Kiani," said he, "is there a woman in it?"

Kiani eyed the mattress; then he looked at Mr. Akeley in earnest reassurance. "No, sare—no woman in it," said he.

"Good," said Mr. Akeley, relievedly.

Ultimately we understood that mattress. Chickens were our daily repast and the mattress was augmented daily by a steady increase of chicken feathers. That explained, too, the featherless neatness of the kitchen which I had commended.

With this cook we had a trifle of French for medium and a pair of English words and one or two Swahili. Our chief resource in interpreting were the Elizabethville boys, Kiani and Mablanga. They knew a little English, Mablanga more than he vaunted, and Kiani much less than he tried to have it appear, but without them we should have been pretty helpless. Kiani had a mediumistic method; he got a word or two as clue and then tried to read your mind. You either wanted a thing or you didn't want it, and so you either got it or you didn't get it. There were happy occasions when your desire and his interpretation coincided. Among the boys he was a humorist and his high-pitched laugh rang out continually but with us his humor was of a more unconscious kind.

"Kiani," said Mr. Bradley one day, surveying suspiciously the water brought for tea making, "Kiani, cook wash in this water?"

"Yes, sare," said Kiani, emphatically, "cook wash him very clean, sare."

Merrick, the third of the Elizabethville trio, was the wash boy, and a very black, smiling-toothed youngster, lazy beyond all competition. Merrick always felt that he had done enough if he brought himself into camp; to do any work after he got in was an infliction of the inscrutable white man's providence. He would accept the bundle of washing and the allotted soap, and then, at every little addition to the washing, the burden of the world deepened upon him. . . . It was just so much time taken from his beloved little musical instrument which he kept strumming between exertions. Our conversation was generally confined to "Pana mazuri" (no good) on our part, anent his efforts, and "Table-cloth pana mazuri," on his. Later on, I regret to say, Merrick fell under a deserved suspicion of theft. In camp a boy might pilfer a little soap or sugar but our belongings and our money we always felt were absolutely safe.

Mablanga was the real backbone of the establishment. He was a very dark boy of a Congolese tribe found East of Elizabethville; he was Kiani's brother, but assuredly of a very different mother. His age was about twenty-two for he figured that he was near fourteen at the time of the Great War. That boy had every desirable and reliable quality. He was quiet and respectful, and amazingly efficient; he cooked and served excellently; he was tidy and loved to keep things so; he had lived with two English families and longed to see England and America and the world beyond his horizon. He never joined the other boys' dances; the night after an elephant was killed when the boys were swaying and stamping and clapping, Mablanga was putting the dining room tent in order, doing the work of three. To Alice he was devotion itself, and he had a sweet and gentle smile for her that showed the boy's real affection. I felt that Alice was as safe with him as with one of us. She thought everything of him, and on the march I would hear her little voice, "Mablanga, in America we have very different houses—houses with steps inside them, Mablanga," and Mablanga's low, sometimes quietly amused, "Yess, missy." He announced that he would not stay in Africa after we left but would always accompany us. "No leave baby, missy," was his continual assertion. I was sorry that at the end we were forced to decide against taking him.

Our relation to these boys in those months of safari was very much, I imagine, the relation that existed before the Civil War between kindly owners and their black possessions. These boys were at our service for any one of the twenty-four hours, for just as much service as we could get out of them. But they were amply safeguarded from overwork—excepting always Mablanga the Untiring—by their native leisureliness and lack of responsibility. There were so many of them that they had much more social variety than we, and I am sure that we afforded them much more interest and amusement than we could get from them.

For each boy we had a book in which Miss Miller, as secretary, put down the time of his engagement and the money he was to receive and entered every drawing he

LEO AND THE UNION SUIT

[page 166]

THE THREE-STONE KITCHEN AND ITS AIDS

[page 166]

The Ruchuru River

[page 175]

Killed by Lion—Grave of Mr. Foster

[page 175]

made against that. On the march they drew out but little, but at the time we paid off the porters at Lulenga, Kiani came to her and wanted all the wages due him—a hundred francs. It was a nuisance to get the money box out and unnailed again, but we did, and gave him his hundred franc note, and Martha made the entry in his book. Then he promptly tendered it back. He wanted her to keep it for him. He had no use for the money. All he had craved was the sensation of being paid.

They came to us for everything—clothes, thread, cigarettes, and medicine. It was a dull day when no boy had a thorn or scratch requiring plaster and bandages. They were very much a part of our lives—cheerful, grinning, tattered companions of the march and camp—black shadows of many a mile—unfailing apparitions to the evoking shout of "Boy!"

CHAPTER XIII

THE LION THAT CAME TO LIFE

My Daylight Lion on the Ruindi Plains

Once the gorilla hunt was over we had planned to turn our attention to some of the smaller fry of the African fauna which usually lure the adventurous overseas—the elephants, the buffalo, and the lion.

We were not at all bloodthirsty, and we hadn't the slightest desire for indiscriminate slaughtering, but we did feel the lure of big game hunting, and I was convinced that I was offering any of the animals I mentioned a more than sporting chance in the present state of my shooting. I had shot at one elephant, two targets, and three crocodiles in the three months in Africa.

Mr. Bradley particularly wanted a buffalo, Miss Miller and I were eager for lions. Miss Miller already had one elephant to her credit and I was hoping for similar luck. We had very little time, for, although Mr. Bradley and I were unhurried, Mr. Akeley had lecture engagements in America, and the constant delays of safari had reduced our hunting plans.

The volcano had cut ten days from the schedule, but the gorillas and the volcano were the high spots of the expedition. Anything else was an after-climax. Still, a lion can be a vivid after-climax.

There is distinct difference of opinion among hunters

as to which is the most dangerous, the lion, the elephant, the buffalo, or the rhinoceros. Drummond puts the rhino first, the lion second; but Mr. Akeley has discredited the dangers of the rhino, believing most of his so-called charges are simply blundering rushes, not actuated by any sight of the enemy.

Mr. Akeley puts the buffalo first, with the elephant a close second, yet he said he would rather hunt elephant than lion; he knew he could stop an elephant—he demonstrated that the day on the plains—but that Leslie Tarlton's experience had shown that a charging lion could come fifty yards with a bullet in his heart.

Stigand puts the lion first, the buffalo last. Frederick Selous, mightiest of big game hunters, puts the lion first and the buffalo and elephant on a par. Colonel Roosevelt stated that the weight of opinion among those best fitted to judge was that the lion was the most formidable opponent of the hunter under ordinary conditions.

In the Congo we had not been in game country to any extent, so we accumulated few stories until we reached Kivu, and most of these were about leopards. We were warned to close our tents and never stir without a light at night for fear of prowlers, and the native runners were never sent alone, but always in pairs.

If half the native stories were true, the leopards exacted an amazing toll. Even allowing for exaggeration, their terror was so real that it must have a good basis of fact. The Belgian officials and missionaries had many instances. At Lulenga the Father Superior pointed to a banana grove that we were passing one day and remarked casually that there a leopard had eaten a young native woman about two months before. The beast had entered the hut at night, seized her, and dragged her into the banana grove. The natives had not attempted a rescue, but at daybreak had gone for the White Father, who came down with his gun, but the leopard made off and there was nothing left of the poor woman but evidence.

At the Mission was a child with fresh leopard marks on its forehead. The leopard had entered the hut, not for the child, but for something else, a dog, I believe—a leopard delicacy—and had wounded the child in its spring.

It is rarely one sees a leopard. They are too wily and catlike. Men have lived in Africa for years without a glimpse of one. Most of those that are killed are got by gun traps at night. On the other hand, Monsieur Flamand of Ruindi, ran into three one afternoon and shot two.

For months we had heard of the Ruindi plains as one of the richest game fields left in Africa, where we could find antelopes by the thousand, and buffalo, elephants, and lions everywhere. There were no rhinos, but we were not after rhino. It seemed the very place for our needs and we planned to move out there with all speed.

Moving with all speed in the Congo means going into camp and waiting for porters. Porters from Lulenga would go no further than Ruchuru, two days away, and at Ruchuru we would have to get fresh porters to take us out on the Ruindi plains, a three days' march. There we would have to get porters from Luofu, two days farther on, because the Ruchuru porters would not remain on the plains.

With real regret we bade farewell to our good friends at the Mission, and on December 4 started north through the Rift Valley to Ruchuru.

As usual, we had sent a runner ahead to notify the chiefs, and at noon of the first day we found a chief out to greet us, with the grass cut for a camping place and eggs and delicious bananas for a present. I spent that afternoon writing on the Corona with the usual crowd of natives sitting curiously about; they believed the typewriter some sort of musical instrument, and must have marveled at the monotony of the air.

Next morning we made a leisurely departure at seven-thirty, and about noon we crossed the wild-rushing Ruchuru River on a picturesque bridge and wound up the slopes into Ruchuru, one of the most important stations on the Eastern frontier of the Congo. It is on high ground, at an elevation of about five thousand feet, with Lake Edward on the north and the M'fumbiro volcanoes to the south. From the elevation the climate ought to be very healthy, but it is not considered as excellent as Kivu. Its wide, spacious, flower-bordered avenues gave it the air of being quite a place. Like Kissenyi and Albertville, it seemed a stage setting waiting to be filled with the actors.

There were several officials at Ruchuru, the Commissioner, the *chef de poste,* the *agent territorial,* a banker, and there had been a doctor, but he had left to escort Madame Deriddar home, leaving the infirmary in charge of native orderlies. It was here that Monsieur Deriddar had been brought and had died and been buried. A domestic touch was given to the streets by seeing two Belgian babies being wheeled about in improvised perambulators, one by a black boy nurse, one by a black girl. It was the first time since entering the Congo that we had seen a native woman employed by "Europeans"—the term for all the whites. Always it is the men, who are called "boys" to their last days, who are employed for cooking, baby tending, sewing, and washing. Alice was enchanted with the babies; the little blue-eyed Elizabeth Piquard was her especial delight.

Here we camped in a grassy square, a former market-place, that became a muddy square through the rains, waiting our porters and making arrangements to leave Alice, for we heard that the Ruindi was too hot for her. Commissioner Van de Ghinste and his bride offered to care for her, but we borrowed the little two-room house that belonged to the Mission Church, hospitably vacated by our very kind Père Van Hoef, who had come up from Lulenga for special services here, and installed her there with Priscilla. They had the devoted and untiring Mablanga for cook and caretaker, Jim for assistant, and all the resources of Ruchuru. Attached to every Belgian post is a herd of cattle, so fresh milk was obtained often from an official's private supply. Filtered water was also provided, and the Mission at Lulenga sent in fruit and vegetables. The thoughtfulness and kindness of the Belgians were unvarying in all these things, which mean so much—especially to the happiness of a child. We went over our supplies here and stored the superfluous things in the warehouse, for we were to return that way, and with seventy porters set out the morning of December 9, northeast toward Lake Albert, with a penciled map from Père Van Hoef to mark the marches and the haunts of game.

Down a long hill we coasted gloriously on our bicycles out onto the plains where a blue rim of mountains seemed pushed back against the horizon. But the bicycling did not last long. The elephant grass was high and the path through it a winding tunnel. Elephant grass is so called because you never find elephants in it; it is an inedible, giant grass, in its earlier stages a flat, sharp-edged, green blade and when matured and dry it is like young bamboo.

At the foot of a long climb we abandoned our bicycles for the following bicycle boys to carry—that is, we believed there were following bicycling boys. But they had abdicated. We never saw them more. They had melted away to their villages, and probably figured among those taken by leopards. It took time for us to be convinced of their departure, and no amount of time reconciled us to it, for we were forced to retrace our steps and push those wheels the rest of the day.

That morning we passed a group of natives in great excitement. They said one of their number had been seized by a leopard in broad daylight, not an hour before our arrival. In the story which reached Ruchuru, the leopard had taken one of our men.

That night and again the next night we camped on the banks of the Ruchuru River, a swiftly rushing stream, whose banks were so netted with tropic luxuriance of palm and vine that we had great difficulty in forcing our way through to see the hippos who were snorting and blowing all about.

Our only way was to follow the hippo trails, which were wide enough, but very low, like tunnels through the thick green, and raked by every imaginable thorn. Possibly the thorn produced an agreeable tickling in the hippo's hide. It did something else to mine.

The hippo comes out at night to feed, and the ground was crisscrossed by these trails. I shall never forget the first hippo I saw climbing out of the water. His back was towards me and I thought that it must be an elephant. They are huge beyond belief.

They were not at all shy, and one old lady who came snorting and blowing down the river with the current, whirled about at sight of us, her eyes rounded, her small ears cocked, her wide mouth set with astonishment. She faced us while the current bore her away, then she sank and either swam back to us under water or galloped on the bottom, and came up nearer than before. She did that half a dozen times, then appeared with a comical replica beside her—her baby—as big as a young bull. She had probably been keeping the baby on her back all the time. They kept coming back delightedly until we wearied of their entertainment and remarking, "kwa heri" (goodbye), strolled away.

The porters wanted one killed for meat, but fortunately for the hippos and our feelings, the animals did not show themselves on land where we could get them. A hippo shot in water sinks and is carried off by the current.

On the second day's march we came to Maji ja Moto,

or Hot Water, streams of really scalding water smelling of sulphuretted hydrogen, gushing from a rock. It was here the Duke of Mecklenburg experienced his most stirring lion hunts, and here a leopard had sprung like a flash at him from a bush, even as he aimed at the gleam of her. His shot had pierced her neck and she rolled dead at his feet. . . .

We, however, saw no leopards and heard no lions that night; we saw only antelope and heard nothing but the incessant snorting of the hippos as they splashed in the river or stumbled in the darkness against our tent ropes.

The third day's march brought us into the territory of the Ruindi. Here Monsieur Flamand, the administrator of the territory and until the last month the only white man within its borders, came striding towards us, having walked a day and a night and half another day from Luofu, his residence beyond the mountains, to greet us.

Our camp on the Ruindi was on a high plateau, cleared like a parade ground for fear of lion and encircled by a moatlike ravine through whose jungles rushed the Ruindi River. Across the ravine we looked out on a wide sweep of plains, with a range of mountains against the sky. It reminded me of some Colorado highland, and we began to wish that we had brought Alice.

Our tents were strung in a military line with the grass rest houses, one of which we used for a dining room. Across from us, on the edge of the plateau, was a mound weighted with stones and embellished with antelopes' horns, the grave of a young Englishman,

R. C. Foster, killed by a lion two years before. He and his brother, both experienced hunters from British East, had gone in long grass after a lion, shoulder to shoulder. The lion had charged, and the older brother's gun missed fire, and the lion seized the young brother by the neck, bit him savagely, then sprang away. The young man lived for two hours.

It was a melancholy reminder with a strong personal interest to those about to seek the lion in its lair. It was here, too, at Ruindi, that Mme. Deriddar had been so savagely mauled. It was this same Monsieur Flamand who had cared for Foster's grave, nursed Mme. Deriddar, and tried to save her husband who had died of the fever.

It was December 12, the day after our arrival, that Monsieur Flamand took us out to tempt Providence and the lions.

Now in British East Africa lion hunting is a ceremony. Half a hundred beaters cover the brush for you, driving the game your way, and you have gun boys who can be relied on in critical moments, and you have horses whose four legs can be used to run the lions down. In the Congo you have no horses, no beaters, and no boys who can do more than carry your guns until you need them. Your only hope is to run on a lion in the daytime, lying up after the night's gorge, or to stay out in the bush all night with a bait to attract lions.

This morning at dawn we struck across the plains, with Monsieur Flamand for guide. The trail ran like a ribbon over the level land to the mountains. Against the dry browns of the burnt grass the green of the scattered thickets stood out darkly.

The plains were simply alive with antelope. There was the graceful Thomas cob, a lovely dun creature, with conspicuous white markings and horns like a big gazelle, moving off at our approach, then pausing to regard us; there was the little reed buck, green-gray, springing away with a sharp whistle of alarm; there were droll, dark topis, slant backed, turning to stare, then with a comical plunge breaking into a gallop and making off in a line like race horses. The topi is a rare antelope, and it was extraordinary to see it in such numbers; it is from four to five feet high, with deeply grooved, backward curving horns. The skin is marvelous, fine and brilliant as watered silk—a dark, warm-toned brown on top, brightening to cinnamon below, with splashes of black on shoulder and thigh, all overlaid with a sheen of bluish gray.

For a few miles we went along the path, then struck out across the plains where the tree clumps and thickets were frequent and might house a lion lying up after the night's hunting. It was necessary to have meat for the men, and Monsieur Flamand brought down a cob buck at about three hundred yards, a perfect shot, but it was sad to see the lovely creature sink down with a swanlike lift of its head. I knew then I should never be able to kill one—and I never did. It was sheer shirking on my part, though, for some one had to kill them for meat and for the skins which we wanted.

At Monsieur Flamand's suggestion we strung out, the better to cover the ground. Mr. Akeley, seeing no signs of lion, went off with Miss Miller to photograph the antelope, but Herbert, Monsieur Flamand, and I walked on, some distance apart, scrutinizing the tangles and little hollows.

"Be careful," Monsieur Flamand had warned me. "Remember Mme. Deriddar."

I wondered, as I strolled along, clasping my gun and eyeing the innocent looking thickets, how I was to be careful and get my lion. It was a lovely morning. The freshness of dawn was in the air, the dew sparkling on the grass. Golden-crested crane preened themselves in a little marsh. There was a pastoral sweetness to the scene, with its grazing herds of antelopes, that was unrelated to anything like lion. I felt a self-conscious humor in my scrutiny of the bushes for the waving tail and pricked ears I had been told to look for.

Lions were unreal—fantastic. But here and there a pile of bleached bones told of their nocturnal suppers. Certainly lions had been here. I had no idea how old the kills were, for the vultures pick the bones in a day.

It was eight o'clock. We had walked a little over two hours. I was just speaking to Monsieur Flamand, whose path had neared mine, when a call came from Herbert: "Here's your lion, Mary—look!"

We whirled about and at our left, a little distance away, were a lion and lioness going through the grass. We raced towards them and they went up a little slope and paused on the crest to look back at us. They were the most marvelous picture of wild life that Africa can give.

For the moment they stood unmoving—green-gray

statues, cast in verdigris. The lion seemed tremendous, his ruffled mane crisp against the sky. Then they made off, with a supple rippling of muscles, heading for a long reach of brush at our right. We ran as hard as we could to cut them off, but they were out of sight among the thickets when we reached the edge of the depression in which the thickets were scattered, surrounded by fairly long grass. The grass was tawny and the thickets rusty leaved, and as we peered and watched we felt the matter of protective coloring was decidedly overdone.

As we stood on one edge, Herbert watched the other side, sure the lions had not escaped that way. The wind was blowing from that direction, so we went around and started a fire to try to drive them out. But the grass burned only a little way, then died at the first green creeper.

We circled the rim, and a native who had joined us told us he had just seen the lions, and in another moment we had a glimpse of the big fellow, slouching across a little opening, his tail waving. We saw the lioness a little farther away, gliding from one thicket to another. They were getting farther off and we were afraid we might lose them, so we went down into the brush, our guns ready, scrutinizing every tangle and tree clump. It was a popular covert. Generous piles of antelope bones told of cob and topi that had been dragged there to conclude an evening of small joy to them.

Monsieur Flamand, advising extreme caution, would seize one of these bones and hurl it nonchalantly into a thicket, observing hopefully, "Now he may be there!" When I did the same, trying to keep my gun pointed at the bombarded thicket, "I have such fear for you—you are so imprudent," he told me. "Remember Mme. Deriddar!"

We went on through the grass, trying to be as sharp-sighted as possible. There is a real thrill to life when you know that at any moment the rusty leaves and the tawny grass may part before a hurtling form. We glimpsed the lioness some distance ahead and started hurrying along on a little trail, Monsieur Flamand ahead and Herbert and I following.

"Keep midway between bushes," Herbert advised. "That lion might be anywhere," and just that moment, eyeing the thicket at my left, I saw the lion. He was about twenty feet away. It was a green thicket, for a wonder, and through a hole I saw the ruddy mane, the ears pricked forward. His right side was towards me, as he was facing the place where the little trail ran towards the thicket.

I did not have time for any fear except the ignoble one that I might not get the lion before the men did. It was to be my lion—they had made me a present of the first one met—but it was so near, so ready to spring, that their guns went up with mine.

I fired as fast as I could, sighting by that ear and neck. Monsieur Flamand fired immediately after, but the lion had gone down at the first shot with a roar that reverberated with the crashing of the guns.

We waited. There was absolute silence. Not a snarl. Holding our guns in readiness we made our way to where we faced the opening in the thicket which the lion had been facing, waiting for us to pass.

A deep growl greeted us and I fired again. There was a gurgling roar that died away to silence, a silence unbroken, even when we tossed in experimental stones.

We tried to wait a decently cautious interval, but we didn't overdo the matter of caution. We went in, guns ready, and there in the cavelike interior a large lion was stretched on his side, motionless, apparently having just breathed his last.

If ever a lion looked dead that one did. He wasn't stirring, he wasn't breathing. We examined the wounds; my bullet had gone in the neck at the base of the skull. The frontal one had gone between the eyes. He had a beautiful skin and mane, and his appearance was majesty itself. I had to remind myself that his death was salvation for three hundred gentle antelope each year.

It was just eight-thirty, and Monsieur Flamand triumphantly reminded me that he had promised me a lion by half past eight. He was certainly a man of his word. Hastily we cut down the overhanging branches and I propped the lion's head in my lap and Herbert took my picture.

Now that picture shows that the lion had his eyes screwed up. If I had known as much about dead lions then as I did a little later I should have known that when they die the eyes are open—not with the lids drawn in a paralyzed sort of wink. The shade was too thick for a good picture and we had the boys, who had now come up, drag him a little bit more into the light. Then I knelt by him for a second picture. In that picture the eyes are wide open—a change I did not happen to notice. I made ready for a third picture by kneeling behind him and trying to hold up the heavy head.

It was a magnificent head. It had a dignity so impressive that somehow it forbade sympathy. A wounded antelope sinking down is poignant, but a lion's life is swift and violent, and a sudden death is no unfit conclusion—it is better than the hyenas.

I was thinking all this and trying to keep the head up when suddenly that lion growled. It was a low gurgling growl, and for a moment I was sympathetic as I held him.

"Is it the death rattle?" I asked. Monsieur Flamand assured me the lion was certainly finished, he wasn't moving, so we went on to get the picture, the lion growling a little more and more. And then—just as that picture was taken—the lion roared! The roar isn't audible in the picture, but it is visible in my expression. It was the most astonished moment of my life. I left the lion, left him abruptly. I joined the men, and we stared at him. We saw that his eyes were open and he was breathing with a regularity and vigor that would have been reassuring to an anxious nurse.

Just then Mr. Akeley, who had heard the shots, came hurrying up with Miss Miller, and I shall never forget the astoundedness of his expression nor the vigorous disapproval of his remarks. Fortunately for the *entente cordiale,* he spoke no French. He was conveying that the lion was very much alive and would recover from his temporary paralysis at any moment and kill us all.

Anthony, as I had christened him—quoting, "I am

Mrs. Bradley and the Lion That Did Not Stay Dead
[page 182]

The "Dead" Lion That Roared as His Picture Was Taken
[page 182]

Topi Staked Out for Lion Bait

[page 193]

Natives on the Ruindi Plains

[page 193]

dying, Egypt, dying"—had ceased growling; he was breathing naturally and his eyes full of intelligence, followed us watchfully. His expression was positively benign, but I did not know how far one could trust to it. So far he had not stirred, and recklessness inspired me, and while Mr. Akeley kept the gun on him, we took one last photograph, and then I shot him through the heart. His convulsive bound before he fell back on his side was proof of astonishing vitality, and then the swift glassy change told me how death really looked.

He was a young male, with a fine skin and mane, the only mane of any lion we got that was not alive with ticks.

The lioness had gone out of sight and, after the grim business of skinning, we went on after her and later walked unending hours across the dry stubble of the plain or through longer tangling grass trying for a lion for Martha or Herbert. We saw three others in the early part of the morning, but they got into wood so dense there was no hope in following.

Later, after a conservative estimate had placed our walk at twenty miles, we felt a positive indifference to finding anything more—unless we found it right in front of us and in a recumbent attitude! We came back to camp after nine hours of exercise and found a herd of elephants squealing and racketing in the woods by the river. They had chosen their time well!

We matabeeshed (rewarded) the gun boys and porters and bathed and changed, and found a black runner had brought in a packet of mail from home. With devouring eagerness we read our three-months-old letters, then dined outdoors in a lovely sunset, while the elephants trumpeted about and the baboons came out on the ravine slopes for food. Then as night came on a fire in the brush spread out in blazing splendor for a mile, flinging the acacias in ebony silhouette. The cracklings came across to us like firecrackers. Then the flame died to an amber glow. The moon swung suddenly out from the clouds overhead and turned the world to silver. Out of the stillness came the hunting grunt of lions on the plains.

CHAPTER XIV

LION HUNTING AT NIGHT

NIGHTS IN A "BOMA"; NELLIE, THE FAITHFUL LIONESS

THE moon was bright now, a tropic moon directly overhead, casting inky pools of shadow beneath our feet. Every night we heard the distant roaring of lions and that hungry hunting grunt of theirs. Their hunting noise is really a series of grunts, beginning with one low one, followed by six or more hurried ones, ending, after an instant's pause, with about three long-drawn-out grunts, generally increasing in loudness. Um—um-um-um-um-um-um——*um*—*um*—UM!

It is almost impossible to tell how far away those grunts are. Whenever you hear them you are always inclined to think the lion is behind the next bush. To hear them from our camp made us eager to spend a night in the brush and see the night life of the jungle and try for lions.

The way to try for lions at night is to kill an antelope about half a mile from where you want to use it and have it dragged that half mile to leave a good trail; then you stake it down in front of the thicket where you conceal yourself and wait.

You try to arrange the scene so the moonlight will be on the bait, with a clear background against which the lion will show up well when he comes to dine. You pile as much fresh brush as you can artistically upon your thicket or "boma," as the hiding place is called, for the lion can see as well by night as by day, and you leave a peephole for outlook and your gun. You have to get into the boma by sundown for the moment the first gray shadows creep over the plain comes that hoarse grunt. And then you sit perfectly still and wait for twelve hours.

Before Monsieur Flamand left us Mr. Bradley went out one night with him, but though they heard lions all night they had no chance at a shot, except at mosquitoes. We had no mosquitoes at all in our camp, but this boma was six or eight miles away from camp, by a waterhole in the brush. However, a waterhole was a good place for game, so the next night Mr. Akeley, Miss Miller, and we two Bradleys went again to the same boma to try our luck.

It was a natural thicket, reënforced with greenery so the lions wouldn't see us too easily, and the clearing within the thicket was so absurdly spacious that we spread boughs for any weary ones to lie upon during the night.

We had our porters carry out a chop box and blankets to this retreat. Then the porters left and we ensconced ourselves and dined sketchily within it. I had a foreboding at the time that the flavor of green onions and cheese was not likely to help conceal our presence. We were surprised, as always, by the swiftness of the tropic night, for at five-thirty by our watches the sun dropped behind the mountains, and at six o'clock Martha,

Herbert, and I were sitting in the dusk, guns in hand, peering out through loopholes in the boughs at the mournful nose of a poor dead topi.

The moon rose, the dark turned silver clear, the lions grunted and roared but not one came near us. But the mosquitoes came. They came humming in swarms and settled on every vulnerable inch of us. But for the strangeness of the scene, I could have believed myself back in dear old Wisconsin on a June night. But we were handicapped here as we were not in Wisconsin. There we could at least attempt to protect ourselves. Here to lift a hand and deal a slap was to bring reproach upon yourself from your fellow, and equally tormented, hunters. Any noise at any moment might cost a lion. With grim self-control we learned to blot mosquitoes in slow silence.

Mr. Akeley was not trying for a lion. He reposed peacefully most of the time upon the boughs, and from time to time the two weary Bradleys took their turns, but Martha Miller sat unstirring upon her camera box for nine long hours until the moon sank and the morning light began to steal over the far-away mountains.

We decided that the location was not a happy one, and that we would abandon the neighborhood of the waterhole. A day later, Mr. Bradley took out the tiny pup tent, at Mr. Akeley's suggestion, and squeezed it into the center of another thicket some distance away, cut three portholes in the sides for guns and set up three steamer chairs within it which filled it to repletion. Then, leaving a porter to guard the antelope he had killed and staked out in front, he came hurriedly back across the plains for Martha and me.

We dined at five and at five-thirty we started off. Our camp was encircled by such a deep ravine that we had always a sharp scramble down and then a long climb up the perpendicular sides of it before we gained the opposite plains. Three luckless porters carried our bicycles up for us and once on the plains we mounted them and set out along the tiny ribbon of native path running out to the western mountains.

The world was bright with color when we started; we were riding straight into the mountains of azure through fields of grass that seemed like waving golden grain. Here and there the scattered trees and brush glimmered with that lovely green that comes just after the sun has gone.

We cycled as fast as we could but the light went faster. The mountains grew darkly purple, then coldly gray. The grass lost its gleam of gold and became vague and mysterious with blotting shadows of forest reaches. The path ceased to glimmer whitely. The world was spectral gray.

At our left we heard a lion grunt. Um, um-um-um-um-um-um—*um*—*um*—UM!

It was uncannily near. We cycled a little faster, remembering that there was but one gun in the group and that was on Herbert's wheel and difficult of access. There was no use getting it out until we stopped cycling. Martha's wheel and mine were a little short for carrying guns so we had left our Springfields for the porters to bring on, and the porters were trotting after us somewhere out of sight. It was but a few minutes after six but the mountains had shouldered the day so quickly out of the way that it seemed like midnight at home.

The lion was keeping right up with us, grunting away, somewhere out on those plains at our left. It was no use for us to tell each other that he was probably a mile, a half mile, away—he wasn't. And he could see us. Probably on our wheels he took us for a new species of antelope. It seemed to me that our crouching pose, our apparent flight, would just naturally invite him to the chase, and I realized that being on horseback was no bar to an attack by lions; I had heard innumerable stories of men set upon when riding casually homewards. Even with my frail memory for names I could remember that there was a Mr. Pease, a former magistrate in the Transvaal, who had been pulled from his horse by a lion.

I wished I hadn't remembered his name. It seemed to make it more real.

And, then, at our right, another lion grunted. There are people to whom a lion's roar is the most terrifying thing in the world, but I think that a lion's grunt, that businesslike hunting grunt, is the most chilling sound that I have ever heard. And these grunts, hard as they were to place, sounded unmistakably close.

Just why the lion grunts when hunting no one knows. One theory is that it is done to start up the game and I can easily believe it—that sudden reverberating intimation of a lion's presence would send every panic-stricken little hoof flying in revealing clatter. Later, when the lion has his prey located and is ready to strike, there is never a warning noise from him.

We cycled so fast that we soon reached the brush and in a few minutes we thought we had come to the right place to leave the path and our wheels and strike off for the boma. There was a lone tree to our left, the landmark that Herbert was watching for, with a bush at the right, and beyond the scattering of thickets. We abandoned our wheels at the pathside and Herbert got out his gun with a speed considerably accelerated by the increasing concert from the lions, and then we plunged into the grass towards the thicket where the ambush ought to be. Herbert had left a boy to see that nothing made off with the bait before we arrived and we began to whistle to this boy. No boy responded. We tried again. Then we called. He couldn't be asleep —not with those lions about. My immediate theory was that he had been eaten by lions—not the present, grunting lions, but other previous lions, now silent and replete.

We stared about in the darkness at the dimly looming thickets. They were all utterly unidentifiable. Any one of them might be our boma. Any one might be the retreat of some lion lying up during the day and just ready to sally out afresh. I remembered the lions I had seen starting from thickets just two days before on the plains, the day I had killed my lion. And when Herbert began plunging briskly about up to the black thickets to find the dead antelope and the ambushed tent I remonstrated feelingly, while keeping close to him and the protecting gun. I reminded him that the

Foster brothers, back to back with two guns, had not been able to ward off a lion in broad daylight.

But Herbert was not concerned with warding off lions. He was only afraid that our conversation would frighten them away before we found the boma. The boma was apparently a little further on. It was certainly not here.

We stood there in the sinister gray dark with the snuffling grunts coming a little closer all around us. I carried Mr. Akeley's little flashlight worked by a dynamo which threw a spot of light a few feet ahead, but I did not feel that it was bright enough to annoy a lion—it was nothing but a little wink of light. As I stared out in the vague reaches beyond us something moved. It melted quickly from bush to bush.

I said in an extremely flat and quiet tone which I trusted indicated perfect calm and absence of tremor, "There is a lion just ahead of me. Point the gun that way."

And then in a similarly unmoved and casual voice Martha remarked, "There's a lion on this side, too. Better keep the gun circling."

We sounded as if we said, "Why, there's Mrs. Brown-Jones coming for tea. Better put on an extra cup."

"They won't come near while you're talking," said Herbert, and he added in a discouraged way, "Now they probably won't come near us all night."

I thought his discouragement was premature. Those two lions were calling across us to each other in a more and more intimate understanding. I can jest about it now—I could jest about it then—but I had a very perfect understanding of what terror was. It was in the helplessness of it all—not having a gun, not having a revolver, knowing that if anything did come I hadn't the defense of a stricken antelope. And the whole situation was shot through and through with a feeling of responsibility for Martha.

Wholeheartedly I advocated abandonment of the brush and a return to the path. Back we went and as we stood there in the path between the grass fields and a stubble of burnt ground, where a heap of bleached bones glimmered with wan reminder, I felt a surge of remorse. That very morning I had laughed lightly at runners who wanted to go in threes instead of twos—I had been skeptical of their tales of a lion attacking them the previous night and felt they were trying to make a social excursion of their errands.

Now I knew that three runners were none too many. Not three times three. And I knew why they kept such fires at their camps.

I tried a little fire myself but the grass would not burn, so we kept the searchlight swinging and talked in loud, nonchalant tones, while the lions grunted back and forth, apparently heartening each other to have the first go at us. "No, you have the first choice. . . . After you. . . ."

Then we heard another noise, faint, far away. . . . More lions, I thought for a moment, but it was the babble of the porters coming on down the path towards us. Nothing was ever such music to my ears.

There, on those backs, were our guns. And there

in those black heads, was the native geography to lead us to our boma. The men came on, half a dozen of them huddled together in a din of talk. We knew then why porters are so conversational. It is their life insurance policy on lion nights.

Paddling swiftly along they soon reached us, in the lead a wizened old mite, spear in hand, my camera strap across his forehead, my camera bumping his bare back. We clutched our guns and a blessed peace enveloped us. That awful helplessness was gone. We fell in behind the leader and that grizzled mite led us on and on for half an hour, then struck into another stretch of scrub.

There was an extraordinary similarity of marking—the same tree at the left, the same bush at the right, the same scattering of thickets. We went farther and farther along, then the guide gave a hyena call which brought a swift echo from the boy on guard in the ambush.

There was our dead topi and there, in the thicket, our waiting tent. Martha and I squeezed in, thrusting our breakfast box and bags with coats and sweaters under the chairs; Herbert hurriedly dragged the kill to a spot a little more in view, staked it down so no lion could drag it away—and the porters fled.

Before we could get comfortably settled, the lions were roaring around us. We crouched in the chairs trying to peer out the peepholes, which were sagging too much with the weight of the brush piled outside and the drag of our guns in the apertures, and wished that we had necks like flamingoes.

It was seven-twenty when we were inside, black dark,

with the kill scarcely discernible. We heard constant grunting . . . we heard the whistle of a reed buck leaping with alarm, then a hard patter of racing antelopes flying for their lives. A lion must be near. We maintained a positive torture of silent, neck-breaking attention.

The moon slipped out from the clouds at last, a blood-red circle bringing the foreground to something like distinctness. And then I relaxed, leaving the watch to the others. I was at the right end and my view was very circumscribed. Martha was in the middle. The first lion was to be hers, and after she fired the first shot we were all to fire as fast as we could to give him all the lead we had.

I was just resting my neck, which had been curved like an Arab steed, when I heard Martha's gun go off into the stillness like the incarnation of all noise. I got a glimpse then of a huge lion, heavy maned, looming before us at an angle to which I tried in vain to jerk my gun, and that same instant the roar of Herbert's big gun followed Martha's Springfield, and the white smoke from it obscured the scene.

Then was a thud of galloping feet. When a lion stalks in to you he is silence itself, but when he gallops off his padded feet hit the ground as hard as a horse's. And then we heard roars, horrible roars. He was at our left, in the brush, evidently badly wounded. The roars became groans, dying down spasmodically. It is a wretched thing to hear a wounded beast groan, but one can fortify oneself by remembering the violence by which he lives, the ruthless clawing down of antelopes. And to-night, remembering those encircling grunts, those dim shapes gliding from bush to bush, I knew just how the poor antelopes felt, milling about bewilderedly, their noses strained, their flanks quivering with fearful expectation.

We heard fainter and fainter noises from the wounded lion and shrill above them the vixenish yelping of jackals teasing him. Then the groans died away. We pressed Martha's hands excitedly. She had her lion, we felt sure. And he was a big fellow with a fine mane.

This was at nine o'clock. One by one the other hours went by, with lion calls now far, now near. The moonlight was marvelous. We saw jackals steal out to the kill and hyenas and two swift cat-like creatures, either leopards or serval cats. Out in the brush a hyena laughed horribly. All the night noises and all the night life of the jungle went on about us as they had been going on from the beginning of time.

It had a wild fascination that stirred the blood. It was Beauty and Night and Violence. . . . Every time a lion sounded near, excitement held us tense. . . . At twelve I was holding the watch alone and I waked the others silently to see two leopard-like forms running in about the kill. They suddenly fled and then we stiffened.

At our left a lion appeared, walking past the thicket to the kill, as silent as a picture on a screen. He seemed tremendous. The dark outline of him against the moonlit grass and sky was a perfect thing, a great, calm, arrogantly assured presence. Majestic is the word for lions. They appropriate it.

Herbert waited till he had sure aim and shot. The lion never knew what struck him down. The guns drowned his roar.

There is no way to tell of the tense excitement of such a moment. The nerves thrill with the old, primitive passions of the time when life was cast on a die. The countless generations that have stalked their prey stir through us. And the blood that loves to race and thrill and is tired of beating tamely through a safe life quickens to exhilaration.

Night shooting has not the danger of the daytime, for unless the wounded beast happened to charge directly at our ambush we had nothing to fear. But it has the element of uncertainty, and the constant edge of expectation that any moment will come the instant upon which all your chances of success are staked.

We saw other lions that night, but far away. One that we took to be the faithful lioness kept running back and forth in the dim distance and calling her dead mate. She flitted by again in the morning dark, and Herbert risked a sudden shot, but we couldn't see what happened.

About dawn a terrific roaring sounded on our left, and we feared that Martha's lion was coming to life again. Then the roaring grew farther away. As soon as we could see in the dimness, Herbert squeezed out of the ambush and glimpsed a lioness off at the left making into a thicket. He fired, but without result. There was no sign of the wounded lion. He had apparently revived, like Anthony, and made off.

One Night's Kill—Mr. Bradley with Two Lions Shot by Him
[page 198]

Miss Miller and Her Lion

[page 198]

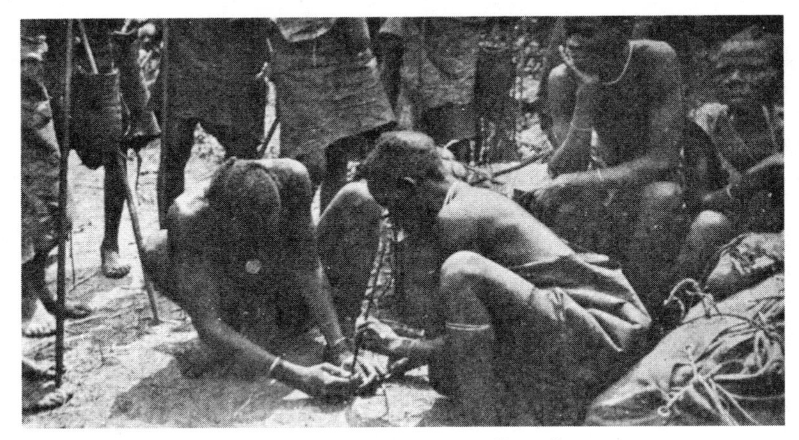

Natives Making Fire with Fire Sticks

[page 220]

Ready for the March

[page 220]

Out in front lay Herbert's lion, a very fine male, and behind him, so directly in line as not to be visible from the thicket, the creamy, soft-furred, graceful creature whom I christened Nellie, the Faithful Lioness.

In each case Herbert's bullet had penetrated the heart, severing the arteries. Death had been instant.

I felt a little sentimental, looking down on Nellie. Her faithfulness had cost her her life, but I was glad that she had gone with her mate and not been left to mourn. . . . But the jungles are ruthless to illusion. Naturalists have no pity. Nell was Nimrod—a young two-year-old male, who had hung about, not from sympathy but for his dinner, and run in for a snatch at the antelope.

Lions are strangely indifferent to other dead lions on the bait. We heard of five that had been shot over the same kill and here we had three that came in, one after another. Several young males often hunt in bands with one or two older, and the apparent lioness that Herbert had seen making off at dawn might have been another such young maneless two-year-old as the erstwhile Nell.

We never did get that first lion of Martha's, although we searched the coverts and watched for days to see if the vultures gathered.

For three more nights we went out again, with one night in between for sleep, but, though the lions roared all about us, not one stepped out where we could get a shot until that third night.

That night the moon was not due until after one. From six to twelve-twenty we had been watching steadily and uselessly through the dark. A lion had been grunting on our right, and apparently coming in towards the kill, but for some time we had not heard him. We had alternated tense concentration with relaxing disappointment.

Then, just at twelve-twenty, from the right came a great shadow, black against the lightening night, still as silence itself. Not a twig cracked under the padded paws. Not a leaf rustled.

It was Martha's lion. I held my gun on him waiting for her to fire first. It seemed an eternity until she shot; she chose her aim with care. Then I followed. There was a tremendous roaring and the lion was down. We saw the great bulk of him in the shadow before us; his roaring filled our ears. He was fatally wounded—Martha's shot had penetrated the brain—but, to prevent an escape and a difficult chase, we flashed the searchlight out one peephole while Herbert gave him the big gun in the heart.

He was a splendid lion. I think he had really the finest mane of all. It was our last lion. Herbert and I went out again and we took out Priscilla Hall, who had reached us now with Alice. We sat up in vain. Though we continued to hear lions every night, we got no more shots at them here—the moon gave us only an hour or two of light now, and the lions were more wary.

CHAPTER XV

ELEPHANTS AND BUFFALO

Big Game Hunting in the Jungles of the Ruindi River

The Ruindi plains, when first penetrated by sportsmen about fifteen years ago, must have been one of the world's richest game fields. The plains lie south of Lake Albert Edward and stretch west to a range of nameless peaks beyond which live the *insoumis,* the natives who refuse to accept the Belgian rule. The level, grassy tract is broken by light acacia growth and occasional patches of brush often thickening into forest, and it is cut by a deep, wide ravine choked with jungle, through which wanders the Ruindi River on its way to the lake.

The plains are simply swarming with antelope, cob, topi, and reed buck, grazing in literal thousands, but the buffaloes, which a few years before had been discovered feeding boldly on the steppe, have been made wary by occasional shooting and have taken to the tangled jungles and the deep marsh of elephant grass on the river edge.

Through those same jungles and adjoining uplands the wide elephant trails tell a story of the past that must have been wealth amazing to the first poachers; now only shreds of the old herds travel those trails, and these are ready, at the first alarm, to stampede back to the impenetrable retreats in the western mountains.

To get an elephant or a buffalo now on the Ruindi means luck and work. We thought that luck had played into our hands the night after I killed the lion which refused to stay dead, for on our return to camp we heard elephants trumpeting on the slopes below.

Mr. Akeley inquired of the natives if the herd were on this side of the river, and the answer, through interpreters, was that it was not. An after examination of the country proved that the interpreter—or the interpreting—was wrong, and that the elephants had been just below us.

If we had known it, we would certainly have gone down after them, amply exercised as we were after twenty miles of lion searching, but, believing that we should have to make a great circle and cross the river to get them, we let the occasion go by and dined with their squeals and trumpetings in our ears.

We planned to go down on them next morning, and Monsieur Flamand had natives out to locate the herd. We arose at dawn and after breakfast started off after the guides, crossing the deep ravine that encircled the camp and climbing up to the plains on the north and east of the river.

There were plenty of baboons in this ravine, and we met them nearly every day scurrying about the slopes or crossing a log bridge of their own. They were aloof but not particularly intimidated and on one occasion a huge fellow remained in the path in front of Herbert and myself, with a menacing air that made us look about for our guns. The gun boys, with the light-hearted lack of responsibility which made their coöperation so desultory, had chosen that moment for a merry bath in the river and our guns were somewhere along the shore. By the time we had snatched them up and advanced again the big baboon had decided not to dispute our way and went off across the log bridge with very grumpy barks. One of them who felt mischievously inclined would make a wicked antagonist for an unarmed man.

This morning we followed the edge of the plains, skirting the canyon through which the river wandered unseen in its jungles. We looked down on bits of forest that were marvelously beautiful and utterly unlike anything else that we had seen. Instead of the crooked, flat-topped trees, so characteristic of African landscape, instead of the low-branching, moss-laden giants of the gorilla forest, we had trees that were fantastically tall and straight, and the effect of hundreds of them in a luxuriance of festooning vine was of indescribable grace.

Among these towering varieties were the gigantic false fig trees. From the bark of this tree the natives make bark cloth. They cut out a piece of bark from the living tree and hammer it with scored ivory tools until it has stretched to surprising dimensions and worked to suppleness. It is very strong when pulled against the grain, but at the least strain in the other direction it shreds like fiber. It has a beautiful reddish color—that is, when first worn. Most of it that our porters wore was indistinguishable from a black rag. The men draped it with classical grace, catching it nonchalantly on the left shoulder.

With a bit of bark cloth, a gourd, a bag knotted of fiber, a clay pipe, a bundle of fire sticks, a knife or spear, all home made, the native is ready for any emergency, and will have warmth and food where your white man could die for lack of matches and tinned meat.

The fire sticks are a precious possession and are usually carried within four or five tightly fitting cases of skin to insure dryness. A hole is cut in a bit of soft wood and into that a piece of hard wood is fitted and twirled about with a rapid, downward motion of the hands. I have seen a skillful man have the soft wood fired in three trials. Sometimes a bit of tinder or charred rag is kept at the junction of the sticks to catch the spark, but the Ruindi natives made fire with only the wood itself.

We walked along the edge until the rim began to break down into slopes and brush covered declivities, leading into the jungles. We passed water buck and topi and cob, who moved off and turned to stare in big-eyed surprise, and at a distance we saw the grasses close over the flitting shapes of a lion and lioness going home after a night's hunting. Our guides began to lead us to the left, into brush that grew thicker and thicker, motioning us to extreme caution and silence in our advance.

The herd was in the jungle ahead. The wind was blowing from us to them—a circumstance that your Congo native takes into no account at all. Mr. Akeley felt decidedly that the element of surprise was going to be lacking in our attack. He was anxious to get motion pictures of elephants before we tried for our hunting, but pictures could be obtained only by maneuvering about with the greatest regard to wind and direction, so he merely followed along with his gun in his hands and his camera men at his heels.

Monsieur Flamand led with Herbert following, and I came after with Martha next and then Mr. Akeley.

The trail grew fresher and fresher; here several elephants had been milling around; here one had stood under a spreading tree. Branches were torn off, bark peeled, a tusk rubbed against the trunk. . . . Now the green grew denser. It was difficult to see Herbert's figure ahead; I had to walk fast to keep up with the men. The difficulty of walking fast over tangled ground, of trying not to let a twig crack beneath you, and at the same time keeping a keen outlook for a shadowy spot of elephant showing through the leafy screens is no mean task for an amateur.

Suddenly everything happened—it happened all at once The brush came alive and snapped and cracked in every direction. It was full of thrashing, unseen shapes. A gun behind me boomed—Martha's Springfield. I whirled about and saw Mr. Akeley's gun go up and off. And at the same time I saw what he was firing at—something big and black opening over the bushes at us—the outstanding ears of an enraged cow elephant swinging in on us.

I didn't have time to revel in the sensation of an elephant charge. I merely had a feeling of blank surprise as if I had seen a large umbrella open over my head. Then the umbrella collapsed. Mr. Akeley's shot directed through the ear into the brain brought her down in her tracks.

Back of them loomed the back of another elephant, a cow, and I fired at her, Martha joining, and after a wavering moment she turned and crashed off. Ahead of me a cannonading told that Herbert and Flamand were having their excitements. Flamand was shooting at an elephant ahead of him; Herbert turned an old bull with a shot in the head. He saw him throw up his trunk, spread his ears and make off. The fall of the cow, who was apparently the leader of the stampede, turned the herd.

We had the glimpse of a baby elephant, a streak of black running away from the dead cow. For some time that little one kept thrashing and blundering about the brush.

We had been coming straight down the trail which was the elephants' way out from that jungle, and, with the wind at our backs, they must have known for an hour that we were on the way. We concluded that all the elephants had bolted and that the hunt was over for the day. The bull that Herbert had fired at had small ivory, so there was no use in pursuing that one, and we certainly did not want to kill another cow.

We went over to the dead elephant and found her a splendid animal with a very noble looking head. She had also a tail with more hairs and longer hairs than Mr. Akeley had ever seen. I measured one—twenty-one inches long. These hairs are black and wiry and very much sought after by the natives for bracelets. By soaking they can be rendered pliable enough to twist and knot. The tusks of the cow could be left till it was easier to cut them out, but the men set to work at once upon the ears and portions of skin that they wanted to save, and Martha and I were sitting down to wait in the shade some distance away, when a thrashing behind us made us reach convulsively for the guns left negligently against a tree.

Either the wounded cow or the *toto*—the baby—was coming back to investigate, but we fired in the air and the noises grew farther and farther away. Once or twice more the elephants came back, and once we saw the toto cruising about in the brush. We kept our guns near while we lunched in a nearby glade.

The marabou storks were winging in from every direction to feast upon the kill and Mr. Akeley remained that afternoon to try for a picture of them. Coming home that night he discovered that our elephants had not all bolted, that the natives had not led us to the main herd. On the slopes below him eleven elephants came shouldering through the green, one after another, and he got the camera in position and photographed them. There was a huge bull among them with splendid ivory. In order to get the last bull to turn and face the camera Mr. Akeley fired across his path and got his picture as he swung around. From the camp we had heard the distant trumpeting, so we hoped for another elephant hunt.

We should have gone that next day. We heard them only one more night. When we went into the jungles again the elephants had bolted, traveling back to the mountains. Only the wounded cow remained, lying up to recover; we heard her trumpet and nearly ran on her once or twice in our buffalo hunts.

The buffaloes kept to the jungles and marshes along the river by day, and in the late afternoon we used to see them coming out to feed on a faraway slope. It was fairly inaccessible and the only way to get a buffalo seemed to be to go in the jungle after it.

That is not recommended, even for veterans. The buffalo is a cunning hunter himself. Although a strict vegetarian—and three of the most dangerous African antagonists, the gorilla, the elephant and the buffalo are vegetarians—he has learned to recognize man as a nondesirable intrusion, and meet his hostilities half way.

A buffalo will charge, not as readily as a lion, but with as devastating an effect. I have talked with a man whose friend was literally stamped into nothingness by such a charge. I heard of several other hunters who had been "blotted" by the buffalo's quiet habit of coming back stealthily, after the herd has gone on, to investigate the alarm.

Altogether Mr. Akeley was reluctant to take in the feminines, so he and Mr. Bradley started alone on December twenty-first. What follows is told by Mr. Bradley, for it was his hunt, not mine.

"We left camp at a quarter of two, crossing the ravine to the plains, then descending to follow the cliff for a mile overlooking the jungles of the Ruindi river. The buffalo had not been seen for several days. In three-quarters of an hour we went down the steep side of the ravine into a gulch-like valley, a jungle of a forest, where we came upon the fresh trail of a single buffalo, following an old, well-worn path.

"The jungle was so dense we could see only a few feet ahead on the path. We took three boys in with us, two to carry our guns and the third for rain coats. After following the trail up the gulch, the wind being towards us, we heard a sound that might be buffalo. We took our guns—the four seventy-fives—from our boys, and made them stay back at a good distance while we crept forward cautiously, step by step, our guns on ready.

"After ten minutes of this Mr. Akeley indicated a spot of gray through the tangled leaves, about fifty to seventy feet ahead. The grayness of the spot made it seem more an elephant than a black buffalo—but the grayness might be dried mud. I looked hard. It was the back we saw. I thought I recognized the switching of his tail.

"It had been decided that if we saw a spot of black, any glimpse at all of buffalo, we were to fire and take our chances about getting the animal or inviting a charge. Maneuvering to a good position for a shot might mean letting the animal get our scent, and this lone bull was one of a wily herd that had been shot over often enough to teach him to take no chances. At the first hint of our presence he would undoubtedly get away without giving us a chance at a shot.

"So now I raised my gun, took aim at that piece of gray, and fired.

"I struck the hip; he went down instantly with a bellow that resounded through the jungle. Mr. Akeley fired right after me. The bull went on bellowing and crashed out of sight among the trees. His noises sounded to me like those of a wounded elephant—I thought for a few minutes that we had run on to elephants instead of buffalo. All I had seen was that patch of gray.

"Hearing a loud crashing and thrashing we knelt in the path, expecting a charge, our guns reloaded and aimed. For some minutes it sounded as if the buffalo were coming back our way, then things quieted. Everything was perfectly still for a few seconds.

"Another crash and another bellow. We waited in this position for fifteen minutes, minutes in which there was not one second of dullness. We kept a keen outlook on every side, for the buffalo will often return by another path to turn the tables on his hunter.

"Presently we heard a succession of roars and then very heavy breathing sounding from a place about twenty feet from the spot where we had first seen the buffalo. We were then sure that he was down and probably fatally wounded.

"We moved a few steps forward and waited cautiously for another ten minutes, all the while hearing the breaking of the bushes, the heavy breathing and the bellows.

"Then we crept up and Mr. Akeley fired another shot at what seemed to be the neck. After a few minutes we took another path, and came in on the same side, and saw a big, black bull buffalo down on his side in a watery marsh. He was very much alive, so I put a shot through the shoulder, and Mr. Akeley followed with one through the back of the neck.

"The old chap straightened out and died immediately. We found that my first shot had broken his right hand thigh, shattering every bone in the leg. It was a fatal shot and he would have bled to death, but he might have held out in strength for a charge, even on three legs, if we had not hastened to finish him.

"He was a big old bull, the scarred veteran of many fights. As head of the herd he had held his position with heavy cost to any who disputed it. He had not gone without punishment himself—his right horn had been broken off and its surface polished smooth, his right eye was gone, and he bore many marks of battle. The head was a magnificent trophy. Under the gray mud his hide was black and heavily haired. I should judge he weighed over two thousand pounds.

"Until one comes into close relation with an African buffalo, one has no conception of the power and size of the great animal. He is beautifully formed; there is clear-limbed grace in his fine lines as well as strength and power; he has virility and combativeness in every suggestion, massivity without awkwardness. The elephant has a certain antediluvian grace; the buffalo has a smartness of finish, an air hostilely alert and inimical.

"It was then three-thirty and it would be two hours before the porters could be brought. I sent a note back to camp to have the porters hurried out, then we set to work at once to skin the head.

"We skinned one side of it, then realized that we

would have to cut the entire head off to get it done before dark. There was no time to wait for the camera. While Mr. Akeley was skinning the head I cut the rest of the skin off, cutting down the inside of the front legs and the back of the hind legs. Before I had gone a foot my knife was dull and I had to sharpen and resharpen every few strokes.

"It was a huge hide, black and thick, not woolly as our American bison, but covered sparsely with coarse hair. I was sorry not to have a photograph of him, but that splendid head, mounted, would be an enduring trophy."

One day after Herbert's encounter with this bull we saw a herd of buffalo come out late in the afternoon on the slopes of the canyon not far away from us. We could see them grazing and strolling about, and with glasses we could observe them minutely. It was a wonderful picture, and Mr. Akeley hurried off with the motion picture camera to circle the plains to a point above them and then descend. We saw him disappear from sight and knew that he was making his way through bush to maneuver for a position, and then suddenly we saw the herd begin to leave.

One by one those buffaloes streaked across the open space and vanished in the forest below, a massive horned bull being the last of all to disappear. Something had given the alarm just as the camera was being put in place, so one of the rarest pictures in the world, superb wild creatures in a rich beauty of background dissolved without a film record, leaving the photographer only the memory and the exercise.

Three times after that the four of us went into the jungle after buffalo, putting in nine hours a day up and down the trails that wound through the network of thickets. We went into marshes where the grass closed over our heads and we had to force our way through the tough stalks; we skirted pools on old hippo runs.

Twice we heard buffalo without seeing them, heard them as they bolted past us, unseen, on trails screened by thickets. Once we thought we saw them when we didn't, and fired—finding we had been cunningly fooled by shadowy leaves. And once—once when we had given up all hope, when we had seen the marks of the herd crossing the river and thought that all the herd had gone and were walking along a path we felt no longer eventful—we came upon them.

I saw the black back through a filigree of trees at my left; I raised my gun and pulled the trigger—and the cartridge failed to explode. For a minute I thought it was a bit of vine that jammed the trigger, and tore at it with a whispered warning to Mr. Akeley, who was just ahead of me.

He had heard the click of my gun, then seen the buffalo. He fired, but just as the herd bolted. I flung out that dud cartridge, then began shooting; so did Martha, who had held her fire to give me first shot, but the buffalo were far on their way. Just one moment of relaxation on the path and one dud cartridge had undone nine hours of rigorous care and stalking.

It was a typical buffalo day.

Nine hours of hard work and one-nalf second of hard luck!

CHAPTER XVI

SANTA IN THE JUNGLE

ALICE HAS A CONGO CHRISTMAS

THE Ruindi plains had been reported so hot and feverish that, as I said, we had left Alice with Miss Hall at Ruchuru, but as soon as we saw the Ruindi we realized that its reputation had merely suffered, like so many other reputations in Africa, and succumbed to slander, and that it was no more unhealthy, at least our particular place in it, than a Colorado plateau.

Hot, it certainly was, hunting at noon under the downpouring rays of the sun a few miles from the equator, hot with a dry, burning, brazen heat that made the dream of tinkling ice in a glass of grape juice—grape juice in any stage—an irresistible mockery and the mention of it a criminal offense; but one could always be comfortable in our tents or grass houses, even at noon, so as Christmas approached and our departure was postponed, we sent a runner in to Commissioner Van de Ghinste, asking him to try to find a white escort to send out with Miss Hall and Alice to reunite the family.

We had hope that the *agent territorial* or the *chef de poste* might be abstracted from his duties, but it was not a strong hope, even for the honorarium we were only too glad to offer, so we were tremendously exhilarated when a runner brought the news that *"la petite"* and Priscilla Hall would be sent out to us with a white escort.

It was a three-day march and we did not expect them until the twenty-second, but on the twenty-first the boys' cry of "Toto hapa!" (baby here!) brought me out of my tent, and there, across the plain, flashed the green and scarlet of the "machila" (the hammock) of the Littlest Explorer. Through the glasses we caught the dark line of the porters and then the gleam of a robe—and there came the White Father, the Father Superior, gallantly walking, in new shoes, too, it developed—with Priscilla mounted upon his donkey—and a donkey is the most priceless animal in the Congo.

In this wise they had made the journey in two days, the only difficulty having been to keep Alice in her hammock. She always loved to walk with the rest, though her small legs could not keep up so hurried a pace, and as fast as we changed porters and languages she invariably learned the new words for "Put me down" and used them busily.

But though Alice made Christmas for us, the problem remained of how to make Christmas for Alice. We were a month behind in our plans. We had expected to reach Nairobi in Kenya Colony, as British East Africa was now named, by Christmas, and now in Nairobi there was Christmas mail and Christmas packages and stores and shops; and here in the heart of Africa we had only two trivial toys—long cherished-for emergencies—and a six-year-old's unfaltering trust in Santa Claus.

Laboriously she spelled out her letters to him, detailing the things she longed for, and dismayedly we read them to each other. Then diplomatically we began to explain; Santa was so far away. . . . Africa had no snow for his sleigh. . . . The packages would be in Nairobi. . . .

But Alice would have none of that. "He'll get as far as he can and then send a runner," she announced, and as our explanations grew more pessimistic, her trust grew more passionately clinging.

"Oh, Mummie, let's hang up stockings anyway—and see what he will do!"

We hung up the stockings and Santa did a great deal. It was the most exciting Christmas the small girl remembered. There was a tree on the breakfast table in the grass house, which the boys had lined with jungle green, a tree trimmed with scarlet kindergarten paper and bright with tallow candles shining away among the piles of packages—and the stocking was bulging.

There was Daddy's long-hidden candy, and there was some of Martha's modeling clay, and Martha's own gold chain and locket, and Mummie's sacred pastels, and home-made painting books and picture books and story books, and ivory bracelets, and elephant's hair bracelets made by Uncle Akeley, and there were native stools and baskets from Priscilla, who had been preparing a most bountiful Christmas for every one—and we all managed to produce and exchange a great many surprising and amusing gifts.

Later another runner came from Santa Claus—this time by way of Luofu, where the Father Superior was now with Monsieur Flamand, the Administrator of the Ruindi, bringing gifts from them both to all of us, rare spears and unusual ornaments and feathers, baskets and bells.

The boys entered heartily into the occasion. Each one of them, down to the cook and the last grinning little helper in a dirty tea towel, presented himself at each tent on the first call for hot water, each with an offering, a cluster of the red-gold flowers of the place thrust in a can or a bottle. By the time we were through dressing every tent looked like a flower show, and the boys were expectantly awaiting the returns.

Previous counsel had decided upon a lump sum to each boy from us all. It was given, and the explanation that it was from us all was harped upon, but the boys promptly pursued every individual of us, yearning volubly for tokens. It was our continual experience that a personal gift, however slight, was much more appreciated by the native than any collective generosity.

We spent Christmas day in a vain nine hours of buffalo hunting. On the way to the jungles, crossing the plains, we had frequent glimpses of wild pig. These were the famous wart hog variety, and anything wartier and hoggier than one of these savage boars cannot be imagined—"Mad Pork," as one Belgian attempting to speak English had called them. Several times we tried to stalk them, but they are wary and trotted off at a gait that without apparent effort took them past running antelope; when they really stretched their short legs and ran they were nothing but a streak of dark. Once I worked my way about a quarter of a mile from bush to bush after four pigs, and then stumbled upon a little reed buck, who sprang off with a whistle of alarm and sprang, of course, in the direction of the pigs. They saw him running, and began to trot, looking cannily back; I risked a shot and the hunt possibilities were over; the pig was about a block ahead of the bullet.

They are great delicacies, esteemed by whites and natives and lions. We had an excellent roast from one Mr. Akeley killed. The warts that give the animal its name were very pronounced in this specimen, two pairs, one below the eye and one about the mouth, and these huge warts, the long-pointed ears, the straggling, bristling bunches of black hairs on top of the head and on the sides of the cheeks and the gleaming white curve of his tusks gave the boar an inimitable air of brisk ferocity. When he goes into a burrow he goes in backwards, so as to present the intimidating tusks to the pursuer. Wild pigs can, however, be tamed, and make pets as safe as gazelles; at Kigoma, on Lake Tanganyika, we had seen two of different species kept as pets. Wilhelmina had pushed her unhandsome nose to one side manipulating the door of her stockade, adding a curious twist to her appearance; when she was drinking her portion of milk from a bottle the effect was enough to make one cross-eyed.

Three days after Christmas, the twenty-eighth of December, we abandoned our hope of another buffalo, and started to march back across the plains to Ruchuru. We had received word from the East Coast, sent in by telegraph and telephone and then by a pair of black runners, that a boat would leave Mombasa, in British East

Africa, some time about the last of January or early in February, and we were trying to make the boat. In the old days, before the war, boats had regular sailings from Mombasa, and could be depended upon to feel some responsibility towards their dates, but now the only way to get out of Africa was to get to the coast near the time when a boat was expected, lay in a supply of literature, and wait until that boat or another put in at the port. This continued the state of suspense in which all traveling was accomplished—it was certainly the Dark Continent in this respect. We knew that, lacking definite information, we did not dare lose a day because that day might cost us a boat.

So we bade farewell to the plains, and took our last look at grazing herds of antelopes, shadowy shapes in the morning gray, their horns glinting in the level rays of morning, and commenced the long trail out of the interior.

Three days brought us back the familiar way through Mai ja Moto, the streams of boiling water, and along the Ruchuru River to Ruchuru, where fresh porters were already waiting to take us out of the Congo. For two days camp was a furious activity of packing, casing the gorilla skins in waxed cloths, and winding the skeletons with straw, listing the trophies and ivory for the obliging customs, and seeing to the sealing of the guns, for we were going east into British territory, where we had no license to hunt.

The Belgian officials were more than kind in helping us off; they were amused at our lack of time, the commodity which is so universal in Africa, but their hospitality did everything in their power. We were very much impressed by the fine type of official we found everywhere in the Congo, by the sense of responsibility towards the native and the seriousness of their colonization. Far from perpetrating atrocities, the Belgian Government is now considered to err on the side of leniency, favoring the black against the white.

Toward us their hospitality was the sort that would have put the early days of the open-hearted West to shame, and never in any dealing, official or business relation, was there the slightest attempt to profit by us.

We had an instance of the scrupulousness of their attitude at Kivu. There we had been allowed the government boat for an excursion to the lava fields if we would pay the price of the gasoline. We paid the presented bill without remark, for the gasoline cost did not seem excessive for the heart of Africa. Nearly two months later the black runner brought in a letter with some hundred francs from Kivu—the *chef de poste* had discovered that the gasoline had been billed to them at a less price and sent on to us our benefit of the reduction.

Our next experience with gasoline, in Uganda, was less heartening.

The last day of the year 1921 was consecrated to packing and good-bys, our last social occasion in the Congo being tea with Madame Van de Ghinste, the wife of the Commissioner. They had a very attractive brick house, reminding me of a bungalow in California, with a lovely veranda whose great charm to us was the lack of screens and glass. Roses bloomed in bowls everywhere about it and the coolness of the shade was delicious after the heat of the afternoon sun in which we had been walking. We found everywhere in Africa that at our altitude a thatched roof gave delicious coolness even in midday heat; we knew nothing there of the sweltering, inescapable heat that American summers can give.

Alice's most poignant farewell was with the Van de Ghinste monkey, little Camembert. At every post or camping place she had found some pet, some Mission cat or native dog or monkey and taken it to her heart and arms, and to all our remonstrances anent certain fleas, "It's worth it," Alice would say stoutly. I could have got her a monkey for her own, but a monkey is a serious responsibility, not to be undertaken lightly by any one who knew Mrs. Akeley's experience with that sensitive, clinging, babylike J. T. of hers. And to any one who has known the little creatures in the freedom of their jungle trees the very sight of one, captive, shivering, apprehensive, chained to some lonely perch, is a sorry thing. Their sensitiveness makes them capable of keen suffering, and no country can call itself civilized that allows one to be jerked about the streets, whipped to perform by some organ grinder, or boxed in tiny cages in a Zoo.

CHAPTER XVII

ACROSS UGANDA

OUT FROM THE INTERIOR; THE END OF A THOUSAND-MILE WALK

NEW YEAR'S morning, 1922, we started to march out of the Congo, leading our two hundred porters east towards the British protectorate of Uganda on a way that could only be vaguely defined by the Belgians. It was the way used by runners with the East Coast Mail, but it happened that not one of our porters had been over it before.

We left Ruchuru on a native path that wound down between banana plantations and then, after half an hour, started to climb on harsh outcroppings of lava rock, over ever higher and higher mountains. Somewhere in those mountains that day we crossed the eastern frontier of the Congo and passed into British Uganda, but when and where it was, we had no idea. We went on forever in search of water by which to camp. The country was wild and solitary, no villages on the way and but very few natives to be met. These few told us that there was water farther on, so on we went, and at last, late in the afternoon, a blue green glimmer in a far-away hollow made us believe that we were approaching a lake, but when we reached it we found that the lake was a field of peas, whose blue green foliage among the darker African green produced an extraordinary illusion of water. On the other side of that field we found our water in a mud hole and our camp by it could only be designated as Somewhere in Uganda.

For the first time in our safari experience water was both scarce and bad. Before we used it that night Herbert himself bailed that mud hole to the bottom, allowing no black within it, and then we boiled the water, as we always did when we did not have spring water, and in addition put in alum. We carried six loads of water on with us, in case conditions were worse ahead.

Next morning we were up by starlight, breakfasted by candlelight and were off at the first rays of the six o'clock sun. Up and down and around the mountains we wound on roads so excellent that our bicycles were a real assistance. A great deal of work was done on those roads during the war under white supervision, and for hundreds of years before that the natives had maintained them, for Uganda had a highly organized government before the advent of the whites and roads were a matter of pride to the old kings.

We met more natives that day, and now instead of shaking hands with a sultan he clapped his hands in apparent delight at sight of us, and any one we met along the way promptly squatted down and clapped vigorously as we went by. As we wobbled along on some of the uneven ridges of the roads to the sound of the clapping, we felt as if we were a trick riding act being applauded and that something in the nature of an encore was expected of us. But it is extraordinary how soon one becomes accustomed to this royal reception. When we met an Arab trader, striding through the country in his white nightgown-like shirt, called Kansu, his goods on the heads of a few blacks trotting after him, and he merely salaamed or said "Jamba" (greeting), we felt defrauded. There was something very soothing in the good old reverential ways.

Inadvertently, we did two days' marches in one that day through abundant misinformation furnished us by one of these traders and some stray natives, and we were ten hours on the march, with lime juice and crackers on the way for lunch. It was astonishing to see how all these irregularities of hours and food agreed with Alice. She throve on them. She had peaked and pined our last year at home on all the sensible routine of her nursery fare. But here, wakened rudely into a cold world at three or four in the morning when we were on march, breakfasting on prunes and a monotony of cereal and tinned milk and tough toast and jam, tucked into a hammock with her doll and a book and water bottle and canvas bag containing a cold lunch—bananas, crackers, chocolate, and chicken—and joggled along all day on the shoulders of the men, dining at whatever hour we dined, four or six or seven, on whatever we had, generally chicken and rice on march and more crackers and jam and cheese, she was as hearty and healthy as a child could be. She never had an instant's illness. Not a cold, not even a sneeze. Life was a continual excursion, an everlasting picnic. She was tired, sometimes; we were all tired sometimes, for from choice we would never have traveled as fast and

THE ACACIA IS FANTASTICALLY FLAT TOPPED

[page 220]

BUYING BARK CLOTH BENEATH THE EUPHORBIA OR CANDELABRA TREE

[page 220]

THE GOLDEN CRESTED KAVIRONDO CRANE

[page 223]

FAMILY SCENE AT LAKE BUNYONI

[page 223]

hard as we did, but with her it was the healthy tiredness of a little gypsy.

We climbed higher and higher among the mountains that day on a road that at times became rudely constructed steps; up and up till space seemed to flow round us like a sea. It was the dry season and the air had lost the crystal clarity of the rains; a haze of blue veiled every perspective. Great mountain ranges lay lightly as gauze against the horizon; azure gleams of rivers and lakes sparkled from the shadowy hollows. Far to the southwest the dim peaks of the Sabinio group and old Visoke lifted above a floating web of gray lace clouds.

We camped that night at Behungi, a mountain top eight thousand feet in elevation, where a clean little grass rest house saved us from putting up the tents, and we slept late next morning—as late as five o'clock. I could remember when five o'clock seemed an hour at which you got up if the house were on fire or the baby cried, but those days were gone forever.

Down went the wide road that morning, and then up again, winding along the mountainside, through forests of bamboo and marshes of papyrus, past marvelously beautiful country, and then the third night we came to Lake Bunyoni.

I still think that Kivu is the most beautiful lake in the world, but if there were another lake more beautiful than Kivu, Bunyoni would be that one. It is a fairy of a lake, with indented shores fringed with papyrus and bamboo and lotus—the only place in Africa where those three meet. In and out the papyrus flashed birds that seemed a thousand colors, blue and purple, gold and scarlet, black and orange; white ducks drifted in flocks upon the water like white clouds; and all along the shore ran the purple line of the lotus of the Nile with fringed petals and a golden heart radiating a heady sweetness.

It was the southern shores of this lake that had been depopulated by the pygmy raids from the forests, and no natives lived there now, but on the western shore, where we were, were clusters of grass huts so picturesquely perched on the highlands that it was hard not to believe in the artistic feeling of the builders. But inaccessibility in case of war had been the inspiration.

We had sent for canoes and the men paddled along the shores sending out ringing calls until a flotilla of twenty-seven was assembled. The canoes were dug out logs, and the paddles had long, straight handles and heart-shaped blades. To maintain your balance in one of those dugouts forbade changing even your mind. We sent most of our goods across the lake that afternoon with a boy to guard them, and then spent the night at the rest house, on a tiny point of land cut across by a line of black euphorbia trees. There was something theatric in the beauty of that camp; those dark dramatic trees against a backdrop of pale, shimmering water had an artistry so palpable, so evocative of mood, that one waited for the tuning of violins and the poignant sweetness of music. Tristan and Isolde . . . Romeo and Juliet. . . .

The next morning was gray and overcast, and our paddlers were so slow in coming that we felt they were going to fail the rendezvous, but eventually the canoes came stealing in through the soft mists and took over the rest of our two hundred loads and all of our two hundred porters. Then we had a desperately hard time getting a tribe whose only knowledge of money was of rupees to accept Belgian Congo francs in payment. We had a row that would have put the Tower of Babel to shame, but finally we lined up the disputants, put the despised francs in each fist and shut the resisting fingers. Then something in the actual feel of those francs operated soothingly and assuaged the racket. Then we tried to buy a paddle, at first in vain, but after one man sold his the rest all crowded around, and we had all that we wanted.

Only one man would have none of the francs. He had otter skins for sale that I wanted for a coat for Alice, really lovely skins, much thicker haired than one would have imagined an otter in that climate to be, and I had struggled to bargain with him the night before. He wanted either rupees or cloth, the white "American" cloth that the Arabs have made fashionable for gowns. I had no rupees and no cloth. I had a French muslin nightgown, but that was too sheer for his needs; he considered it seriously, impressed, hesitant, but judgment went against it; he wanted stouter stuff. So I persuaded him to take his skins and come with us to Kabale, the first English post, where at the administrative Boma of the Englishman I would give him his rupees.

All we could learn of the way was that somewhere after Lake Bunyoni, two days or two hours according to differing informants, we should come to Kabale. For an hour after we left the lotus-fringed shores of the lake, Martha, Herbert and I, and Mablanga with Alice cycled rapidly down a road on which we passed increasing numbers of natives in the white gowns of Arab fashion, and at last, on a height before us, we saw a cluster of roofs and the fluttering colors of the Union Jack. This was Kabale.

On one hill was the Church of England mission, on another the Mission of the White Fathers, and on the elevation between them the government Boma and the bungalows of the officials. Up the hill and into the Boma we marched, dusty, khaki-clad figures, greeted by very astonished, white-clad Englishmen. We had sent a runner ahead asking for porters to be in readiness, but nothing in our mention of a party of six had prepared them for this feminine invasion from the interior of the Congo, and the presence of a little six-year-old girl, carried along on the wheel of her boy with a doll at her side was distinctly unexpected.

Commissioner Adams and Captain Persse divided our party for luncheon and in our dusky khaki and mud-stained boots we entered a bungalow California-deep in roses and sat down to luncheon amid the chintz and china and silver and crystal of old England. Martha, Alice, and I lunched with Commissioner Adams in a home from which the young and lovely mistress had been taken just two weeks before. A little baby boy only two months old was being cared for at the English Mission until the Commissioner's sister could come out from England. At the table with us was the Commissioner of the Province, Mr. Cooper, a thin, bronzed man

Waiting for Canoes on Lake Bunyoni

[page 226]

Native Dugouts on Bunyoni

[page 226]

THE TOMB OF KING MUTESA, IN UGANDA

[page 233]

THE END OF SAFARI

[page 233]

of many years' service; he told me that in England were his motherless twins of five and a half that he had not seen for nearly three years now. He looked at Alice and estimated how his little girl would look. At every post, Belgian or British, that price of empire was brought home to us.

After luncheon we scurried for our tents and hot baths and the white raiment in our "air-tights" and paid off the porters and then began on the formalities of customs inspection. The formalities were solemn ones here; and we made innumerable lists, and estimates of everything left in our supplies to the last prune. At intervals I sent a boy out for the native with the otter skins, who seemed to have melted from Kabale before I could secure my rupees; we discovered that he brought his skins to the Boma, the white man's place of administration, and finding a visiting white man who wanted skins, had promptly sold them and retired. Captain Persse saved the day—and Alice's coat—for he kindly sent out and had other skins brought in for me.

The porters for whom we had sent a runner ahead were ready for us and next morning, after the last customs list was checked over, we sent them off with a headman, and we followed at one o'clock after more sociability. After the first downward plunge of the road we could ride our wheels, and the fact that the march was estimated at from sixteen to twenty-six miles did not daunt us.

For six days we went on good, red gravel roads through a very beautiful country, mountainous though with none of the great single peaks that we had grown accustomed to in our M'fumbiro friends, but a continual succession of ranges and ridges and valleys. The slopes were deforested, firewood was scarce and water was scarcer, and we guarded our precious pails of it in the tents, as the boys used it with the same prodigality as if the land were flowing with it. Their unconcern when they knew the facts in the case as well as we did was a marvelous application of the take-no-thought-for-the-morrow admonition.

At Lubando, where water was scarcest, I saw a little black tent boy, the "boy" of our Jim, calmly empty out one of the two pails of jealously hoarded water in order to use the pail to stand on to arrange the bed nets!

We had heard that there was water five miles away and we thrust that empty pail handle into his hand and sent him forth to verify the rumor.

Natives everywhere had been eager for "dower" (medicine), but on this march they were simply clamorous and we could never get settled in camp an afternoon before the procession of sultans and cohorts and private citizens would begin to trickle in and come to attention before the tents.

"Nataka dower, Mamma" (want medicine, Mother) was the invariable request to me, accompanied by illustrative showings of scratches and ulcers and rubbings of the stomach and various seats of pain. For really serious cases we consulted the medicine book and Mr. Akeley's experience, but in general our remedies did not vary much and our prescription rule was simple—we gave a chief three times as much of anything as we gave a minion.

I drew the line at wives. When they began on the symptoms of the dear ones at home I said "Kwaheri" (Good-by), for the dear ones were unlimited. Cattle and wives were a sultan's wealth; cattle were a luxury, but wives were an asset, for the wives worked.

A thrifty citizen could acquire several wives in the course of time, but a really thriving sultan would have a lot of them scattered about his villages. The customs varied extremely from tribe to tribe.

The great struggle of the missionaries, Protestant and Catholic, is to make the natives accept the monogamous marriage. When they have been successful the superfluous women have created a new problem—and often a new class—abandoned ladies in every sense of the word. Mohammedanism, with its plural marriages, has a tremendous advantage. In the old days the tribal standards of morality were high—at least the execution of the laws was rigorous—but along the white man's way the natives have acquired habits that certainly savor of promiscuity. The boy that in his native village would buy a wife and keep house with her, here strolled into the outposts or the towns, flirted, made indiscriminate and highly successful love, and strolled out again. . . . And sometimes the white man, in the lonelier places, made love. I have seen more than one strange safari streaking its way across the solitudes, unmindful of the glasses trained upon it—porters, cook, boys, a white man in his chair and at the end a flutter of red calico that said, "There she goes." Many an old timer in the wilds, losing the hope of return or the companionship of his own kind, solaces himself with what the savage land offers.

No one will ever write a "Butterfly" about those little girls. They lack the soft, silken stuff of romance. They are impudent little episodes, smart, giggling, unashamed. I don't imagine one ever died of a broken heart. I don't imagine a man would die for love of one. But I knew of a man that married one—he had a quixotic sense of responsibility toward a coming *café au lait* generation—and sent a bullet through his head the year after. I know, too, of other cases where a father has sent his illegitimate mulatto girl to Europe for an education.

Near Kabale I heard a very dramatic story of a native woman. In Rukiga country the native tribes, the Batiga, were overrun by the Baganda, whom they hated. The Batiga had always been divided by petty jealousies and feuds, but at last, in this common hate, they held together. They set a night. That night a knife was to go home through the heart of all the Baganda. Now those Baganda men had Batiga wives, who were the mothers of their children, and those wives knew that plot, but not a woman warned them—not a woman but one.

That one must have sat a long while by her fire watching her sleeping husband. At last she waked him, swore him to secrecy and told. He fled like a shadow down the road to safety, and like another shadow, unseen of him, she followed after. His way went by a hut to which a path branched off. It was his brother's hut, and the man stopped, hesitated, and then swerved towards it. The woman sprang upon him from the shadows and her knife went through his back. There was not a sound to give warning. . . . Him she would have saved, not his people. . . . She sells corn now, a worn old woman of thirty years. . . . Hard to re-create from that flabby flesh the fierce young thing whose swift leap cost her son his father's life.

On January 10 we reached Mbarara, in Ankole, a British post of two families, where we had arranged for automobiles to meet us, sent out from Kampala, a hundred and eighty miles away. Seven cars came for us, trucks and four five-passenger touring cars, Reos, Dodges, and Fords. One good truck could have held our loads, but the bridges were so frail that light loads had to be used.

Here we paid off our porters, the cook and all the tent boys except the three from Elizabethville whom we were to take on with us to the coast and return to Elizabethville via Dar Es Salaam. We saw the long line of blacks start down the way back to the interior with a keen pang of regret. The old days with them were gone. Our walking was over. We had walked a thousand miles in all, marching and hunting, over plains and up and down mountains, and some of those mountain miles, we felt, ought to count as two.

We did the next hundred and eighty miles in two days and then, at Kampala, we waited two weeks for the last car, which had broken down and had to be sent out for and brought in. It was of course the car with the gorilla skins and the things without which we could not leave. At our former pace of fifteen miles a day we could have marched the goods in, in less time—and those seven cars for two days cost us twenty-four hundred dollars.

CHAPTER XVIII

THE TOMB OF KING MUTESA

Sixty Years Ago and Now; the Tourist Trail Again

Kampala, built on seven lovelier hills than Rome, is a thriving place of about two hundred whites and thousands of natives and Indians in the teeming bazaar streets. At Kampala our safari was really over. Here were shops and hotels and clubs and telephones and ice tinkling in tall glasses—that ice we had dreamed about on the Ruindi—here were movies and private dramatics and dinners and dances and everything.

We put away our khaki and hobnails and put on white silks and rode about luxuriously in rickshaws drawn by one boy and pushed by another, on our excursions to the bazaars or to the tops of the neighboring hills. Namirembe Hill is crowned by the English Church Mission and a huge cathedral to accommodate seven thousand is being built there now to replace the old building whose thatched roof and conical towers were a part of the old Uganda landscape. Near by is the famous hospital of Dr. Cook. Nyasemba Hill has St. Joseph's Mission, and Rubaga Hill is occupied by the White Fathers, and another great cathedral is being erected there. On Mengo Hill is the residence of the King of Uganda, a young man, King Daudi, with whom much pains in education have been taken. He lives behind the grass stockade of his fathers, in a bungalow well furnished in English fashion, but surrounded by the native huts of his attendants, thatched and woven in an intricate and beautiful manner. The court of Daudi is merely a flourish of the old tradition. Royalty lies dead in his grandfather Mutesa's tomb on Kasubi Hill.

We made a visit to the tomb which used to be a place of pilgrimage, scrupulously maintained by the natives; now it was the tourist trail, with a caretaker and a price of admission. But it was an empty trail; no Europeans had come for a very long time, we were told. There was a deserted air to the ancient enclosures behind the high stockades woven in the royal pattern; in the last court a sun-baked parade ground stretched to the door of the house which had been a great king's palace; left and right circled the smaller huts that had been the treasure houses or the homes of wives. Everything was vacant now save for a stooping beggar and a naked child or two; only the white-robed caretakers strolled about the empty spaces.

It was interesting to re-create the scene of Mutesa's lifetime; the four inclosures thronged with the courtiers whom etiquette—an etiquette whose infringement was death—required to pay incessant court upon their king. Here were powerful lords and overlords and their attendants, warriors reporting their last raids, the wretched captives in their train, hunters with the spoils of the chase and the animals snared in their nets, fishers with their gifts, potters with their finest jars for tribute, disputing men with cases to be heard, men with violent accusations, accused men haled along in bonds, medicine men, bands of musicians—the flute players, the pea gourd rattlers, the harpists, the drummers,—a jostling throng interspersed with dogs and flocks of goats and sheep offered for commutation of the death penalty—exchanging the death penalty for a fine was a great source of royal revenue—a motley mob threaded by the little pages of the king, who darted like wasps on every errand, not daring to walk lest they should be killed for sloth—a world of fear and hope, pride and ambition awaiting a despot's favor. . . . No naked savages in goat skins these; here were shrewd, ingenious politicians, netted in a mesh of meticulous conventions. The men's care in clothing was extreme; to show an inch of naked leg was disrespect punishable by death. Yet, curiously enough the king's valets were women, utterly unclothed.

Through the babel of this gathering throng would come the roll of the king's great drums announcing his appearance, and the crowd would surge forward, pouring in the inclosures after the lords of Uganda, making their way into the king's presence to prostrate themselves in wriggling admiration and then squat at devout attention in a dense half circle before the sovereign.

The king's reception house was the vast beehivelike building which we were nearing now as we walked across the empty square; it was marvelously woven of the canelike elephant grass, or the tiger grass as it was called here, tied intricately together, with an elaborately thatched roof. Within the doorway the king used to show himself sitting on a blanket spread on a reed throne, by his side his white dog, spear, and woman —the Uganda cognizance, for the father of his line and the conqueror of the country, Kigima, had come with a dog, spear, and woman—some favorite pages in attendance, and behind him, in the back of the hut facing the entrance, a cluster of his women.

It was death for a man to gaze upon the king's women. So the courtiers presented a stooping, slant-eyed, cross-eyed look; it was death to touch the king's robes or throne even inadvertently; no man might stand in the presence while the king was sitting or standing; he must sit lower than the king, on penalty of death. It was death, in fact, for anything or nothing; a favorite wife who had the presumption to offer Mutesa fruit was promptly killed upon the spot.

Caprice and cruelty, a nonchalant barbarity that makes our blood run cold with its hideousness of torture, power, dizzying, absolute, that was Mutesa's existence. No Caligula or Nero could compare with him. He was supreme. No man dared speak unless he spoke. No man dissented. The gay, sunny hill was a shambles and a furnace.

I looked into the interior of the old palace. It was so dark after the blazing sunlight of the court that it took time for my eyes to distinguish the outlines of the dim interior. I saw a perfect forest of ebony poles reaching, like a sea of masts, up to the high roof, leaving a lane to the big block of the tomb itself, which was covered with bright cloth of many colors. Many poles nearest the tomb were wound with bunting and bright cloth. Here, beneath the calico, in a metal-covered coffin, made by the missionary, Mackay, wound with white cloth of honor, lay Mutesa, son of Sunna, seventh heir of his invading line.

Mutesa's was the last great reign. After him came swift confusion and disorder, civil war among his sons, and about them the struggle of rapidly growing power of the invaders, Mohammedans and Christians, Protestants and Catholics, until an Anglo-German treaty assigned Uganda to Great Britain, and in 1893 the British flag flew from the grass peak of the palace.

It was sixty years ago exactly, that the first white man to penetrate these wilds, Captain J. M. Speke, was received here at King Mutesa's court. Sixty years is not a man's lifetime. . . . And now, sixty years after, our small Alice, the first little American to travel this road, was standing in the dark hut, staring at the deserted tomb. Back of us from the shadows came whining supplications and gnarled arms of beggar women held timidly out. . . . On either side, among the forest of ebony columns, we discerned forms crouching on bits of mats. . . .

Sixty years since the first white man. . . . Those crawling creatures with white wisps of hair falling over their wrinkled faces could recall Mutesa in his unbridled power. . . . From this place they had seen his victims dragged off to torture, to dismemberment, to roasting fires. . . . They had seen the proudest chiefs of the kingdom prostrating themselves in sycophantic adoration on the ground. They might well have been present when that first white man strode forward, opening his strange, mirth-provoking umbrella. . . . And now their sunken eyes gazed out with a flash of wonder at the little white girl with the fair curls who came to look at the tomb of their dead king, and then danced so fearlessly out into the spaces where once it had been death to approach.

The administrative center of the Protectorate of Uganda is at Entebbe, on the shores of Victoria Nyanza, twenty-five miles away. We motored to Entebbe in a fraction of a Ford with an Indian driver whose aberrations brought us nearer death than any lion or elephant. It may have been the fervid gratefulness of our escape from his clutches and the miraculous sense of renewed life, but Entebbe seemed to us one of the fairest spots imaginable, a bower of a place on the shores of the great lake, which every writer hastens to speak of as the inland sea of Africa. The word Nyanza means lake or sea or sometimes simply water.

On a slope above the lake was the government arboreum where a great variety of tropic trees was grown. There were palms whose hidden chambers yielded clear water to the traveler in distress; there were huge incense trees with sweet scented gum oozing through the wounded bark; there were groves of chocolate trees with hard, dark, shining fruit, like an oval orange in shape; there were rubber trees where from a tiny cut the pale, strange fluid stole out . . . to be rolled into a true rubber ball.

In and out the trees, over the shores of the lake, there darted a winged multitude of birds, glittering little sun birds, preening and darting, golden weaver birds who build huge, hanging tenements of community nests, shrikes and thrushes, finches and warblers, and many, many others for which we had no name.

Africa had given us a wealth of bird life from the Lualaba River days when the black ibis and white egrets had streamed like raying clouds before our boat; we had seen blue heron and white heron, bustard and ostrich; storks like picture postcards of Holland; and the Egyptian goose by its low-hanging nests through which the young drop into the water when hatched; bulbuls and cuckoos and bell birds had been the music of the silent places. We had watched the white Battleor eagle on his lonely flights; we had heard at many nightfalls the hoarse calling of the Kavirondo cranes as they winged their way past camp, and in sunny marshes we had come upon the gleaming beauty of them, their velvet black and whiteness crowned with its radiant golden crest. We had known the marabou stork with his precious fluff of white feathers beneath the bronzed blue-green sheen of his broad feathered tail; we had glimpsed the jeweled blue and crimson of gorgeous plantain eaters and the demure gray of parrots. We had grown familiar with the great vultures who came streaming to roost above our kills; and our intimates were the handsome, white-necked black crows who frequented our kitchen and the little black and white nameless fellows who walked so fearlessly about us at every camp.

We felt that we were just beginning to make many interesting acquaintances, and now it was already growing time to say good-bye to them.

We had luncheon and tea and "Sundown" as the guests of Dr. Fiske, an American physician for many years in the service of the British government here. Dr. Fiske has a tremendously interesting work before him now, the repopulation of the Sese Islands. This group of lovely islands, about sixteen in all, was so ravaged by the sleeping sickness that ten years ago the British government removed every surviving native and isolated the islands, to stamp out the appalling plague. Now the infection is gone and the remaining natives are being taken back to their former haunts. "Each man," said Dr. Fiske, "wanted exactly the place he had before; if he tilled a zigzag plot on an inconvenient hill he wanted that precise field, not the one nearer or richer." In many cases it was difficult to determine the old boundaries for much of the vegetation had disappeared. In places it had been cut down to remove the shade which was so favorable to the deadly tsetse fly, and in other places the antelopes had devoured it.

In those ten years that the Sese Islands have been given back to an utterly manless nature, an interesting fact about the situtunga antelope has been revealed. This antelope, which is closely related to the bush buck, but larger, with long hoofs and shaggy hair, has been considered by naturalists exclusively a creature of the marshes, making its home in thick, reed beds. It was very shy and wary and few white men have shot or even seen it. There were several living in the marshes on the Sese Islands. Now in the ten years of utter peace, the situtunga has come out of the marshes and frequents the highlands, grazing in flocks like any plains antelope—a swift reversion to what must have been its natural environment before the pressure of danger drove it timidly into the swamps for refuge.

No cure has yet been found for the victim of the sleeping sickness, though physicians have been busy with it ever since the sickness came out from the dark forests of the Congo like a blight upon the dwellers on these peaceful shores. Infected areas can be controlled and isolated by cutting down the shade in which the fly loves to dwell and planting the citronella grass to which it is so averse; and by removing all the population, as was done in the Sese Islands, the disease will die out for want of reinfection.

The fly is a sinister thing, longer and narrower than a bluebottle, gray and black with crossed wings. We had seen several on the Lualaba River in places that had formerly been sickness areas but which were now considered safe, and while these flies were undoubtedly uninfected, they were distinctly disquieting things to have about. It has a near relative of similar appearance, which is as deadly to domestic animals as this one is to human beings. When the human biting fly does not have man to live upon it is supposed to subsist on the blood of crocodiles. It was at Entebbe that I first heard the news that a cure for leprosy had at last been discovered and that lepers were actually being discharged in Hawaii from Molokai Island. . . . What a day for Africa when a remedy is found for the deadly plague of sleeping sickness! One bite of a swift darting fly and the victim is doomed to lingering torment and certain death.

After the Entebbe trip, life in Kampala for Mr. Akeley was a pleasant memory of the peaceful hospital to which he betook his malaria while waiting reports of the missing gorilla skins. It was more varied for the rest of us.

There was a daily descent upon the Trading Company, with the invariable report that the car sent back for the other car had also broken down but that a new man was going out that very day; there were constant descents upon Smith, Mackenzie & Co., Limited, the shipping agents, who knew nothing of the possibilities of a steamer but did know that the Union Castle steamer, advertised for the end of February, was not coming for another month at least; there were frantic interchanges of telegrams with the agents in Mombasa to discover the whereabouts of our trunks, shipped three months before from Kigoma in Tanganyika to Dar Es Salaam to meet us at Mombasa. We appeared to have neither ships nor clothes in which to return. Being safely in Africa there seemed every reason to remain there, and I admit that but for certain necessities at home we should have liked nothing better for some time to come.

Kampala for Herbert meant the multitudinous details of disbanding the equipment and selling the tents, bicycles, etc., and the packing of our various accumulations. As he wrestled with baskets and tusks and elephant feet, and the disposition of spears and gourds, and the problems of poisoning skins for their incarceration in a ship's hold for a long, hot voyage over seas; and with getting men to agree to make boxes and getting the men to make the boxes, and getting the things into the boxes, and getting the boxes into the train on the ultimate afternoon when the skins finally came in and we raced to catch the weekly steamer, his ideas for the equipment of his next African trip dwindled to the simplicities of the native, and his next collection of trophies, he resolved, would be of the dodo or the unicorn. Nothing less remarkable was ever going to be packed. . . .

The great difficulty of getting anything done in the tropics is that the working day is so short. The day itself begins early enough, heaven knows. It begins at five-thirty with a cup of tea brought to your bedside.

I loathed tea at five-thirty. I loathed five-thirty. After the strenuousness of safari I wanted sleep—lazy, comfortable sleep stretching on into the forenoon—say until seven. I told the hotel boy about myself and my desires in explicit Swahili. When he brought the tea the next morning at five-thirty I sat up and said, "Pana chi" (No tea), in fierce rejection and that night I locked my door.

"Now!" said I, triumphantly.

But no. Five-thirty—and a tentative trying of the handle. I lay still. A knock. A louder knock. A rattle. Then receding, frustrate steps. . . . I smiled. . . . I prepared to drift beatifically off—and then a click from the French windows opening on the verandas and a white clad form slipped stealthily in, approached my bedside with pantherine celerity and deposited in triumph the tray of tea. Honor had been vindicated. His job was to bring me tea and he brought it.

I succumbed. The motto of that hotel was Service, and except in the matter of the tea we appreciated it. It was the most comfortable hotel that I ever found in Africa, or in America for that matter, and we have most pleasant memories of its spacious, airy rooms, the cool verandas, the delightful meals, and the services of the boys who paddle about like barefoot, white-robed, ministering angels. The name of that hotel is the Imperial and I want to record it.

The tea was only the first intrusion. It was all very well to ignore its steaming presence and drift off to sleep, but the sleep was momentary—another knock preluded the entrance of the boy to take your shoes to clean and after that the boy came back for the tea cups and the shoe boy came back with the shoes and the bath boy came with the bath. We had kept our three Elizabethville boys and found them useful for errands and for our washing and pressing. The verandas outside all the bedrooms were often a scene of great domestic activity, a big black boy ironing away on a little table or sitting cross-legged sewing expertly. The flat iron of Africa has a hollow filled with glowing charcoal so it could be used anywhere.

The house boys of East Africa wore invariably the Arab robe falling to their feet and the small embroidered cap of Zanzibar, while many natives we saw on the streets aspired to the white man's "shorts" and a pukka shirt and—last flight of opulence and aristocracy—shoes and puttees. Often they wore the puttees without the shoes. Our own boys who had come in the bizarre tatters of safari now beggared themselves for white shirts and shorts made by the Indian tailors.

After the advent of bath you resigned yourself and got up as you ought to do to enjoy the lovely freshness of morning. After breakfast we used to stroll out on the veranda to see what the traders had brought for the day. Under a shady tree the venders spread their wares —leopard skins and bowls of black Uganda pottery, the bark cloth so abundant in the region, the typical tall drums, brightly colored baskets from the Soudan, and beautifully made musical instruments with sounding boards covered with lizard skin. . . . Only Herbert was sternly cold to these enticements, and when we yielded to bowls and baskets we approached him with the qualms of the confessional.

The business life of the tropics begins about ten, the social life at eleven when morning calls and lemon squash or anything else are vogue; from twelve to two siesta reigns, and then business matters may be transacted until four-thirty when the sacred rite of tea concludes the strenuous day. The waning hours of sun, the helmetless hours, are consecrated to golf and tennis, and at six-thirty the world gathers for its "Sundown." That ceremony needs no explanation to any old travelers in the tropics; to explain it to an American audience would only wake the regrettable resentment against Law Enforcement too often noted in these United States. The rite, however, is often celebrated in ginger ale or lemon squash—but it is always celebrated.

After sundown the social drift is towards the club,

and after dinner to the club again unless there is a dance somewhere else. Between packing and sociability and reports to the invalid at the hospital, life was varied at Kampala. Major F. A. Flint was our frequent host a tea, sundown, and dinner, and we shall not soon forget his heartening help in any difficulty and his fund of stories—the generous way he detached any admired trophy from his walls and sent it over on his boy's head next morning made us hesitate to utter a word of admiration. From him we secured the rare situtunga horns and a pair of the wide-branching horns of the Ankole cattle.

CHAPTER XIX

GOOD-BY TO AFRICA

Victoria Nyanza to Nairobi; Mombasa and the East Coast

January 24 saw us off at last, by train, to Fort Bell seven miles away and then on the *Rusinga,* an attractive steamer, across the serene Victoria Nyanza, blue and calm and smiling about us. The low-lying wooded shores were very different from our memories of Tanganyika and Kivu. They were charming but we missed our mountains.

There was a stop at Jinja, a breakfast with the Commissioner and his wife, Mr. and Mrs. Henry, and a rickshaw ride to Ripon Falls where the waters of the Nile take their first plunge from out the lake. The cascades are not tremendous but they are very lovely; a little cormorant-studded island around which we saw otters swimming gives picturesqueness, and the first glassy flow of the deep waters over the submerged rocks has unfailing fascination.

This was the tourist trail again, and here were tourists from Boston, two enterprising ladies who had been told by Messrs. Cook of Cairo that they could come into Africa as far as the Birth of the Nile without setting foot to the ground. Here was the Birth of the Nile and here they were, and I gathered that set foot to ground they had not. Judging from the sing-song plaints of the riskshaw boys who pushed them away, a little exercise would not have been unappreciated.

Off across the Nyanza again, and at noon of the third day we reached Kisumu or Port Florence in the Kavirondo country and changed to the train for the high climb to Nairobi.

We woke to see a band of zebras staring at us, so close we could fairly see their eyelashes—fat, chunky little fellows like painted polo ponies. Ever since *Swiss Family Robinson* days, I had been nurtured in the belief that the zebra was fleeter than any horse. It was pain to learn that a good horse could run one down. All day we went through a beautiful country, vast plains alternating with woods of the Australian eucalyptus, planted in close packed rows. At intervals we stopped at stations for breakfast, luncheon, and tea. In the plains we saw our first kongoni; he was a quizzical looking old antelope, his long face of bland curiosity keeping watch over any particular friends, antelopes or zebras, grazing at hand. We passed Naivasha, a lovely lake with a floating island. Hippos abound in its waters.

The country grew flatter and flatter. Suddenly tin houses appeared—the little corrugated shacks of our west. The train ran through a cluster of them and came to a stop in a station of hurry and bustle, stalls and papers and magazines—Nairobi. Nairobi, whose wide street of shops and rickshaws had so foreign a flavor to us, during our week of waiting at the Norfolk hotel, but where the old-timers gather at that Norfolk bar and

"Them days is gone forever," is the burden of their song. Nairobi of stone country houses and private racing stables,—and tea and terraces and sunken gardens and dinners and dances: Nairobi, where a motor's ride away lions are roaring and the glistening Kikuyus, naked but for leaves and paint, are dancing beneath their sacred tree.

We met there many whose names are familiar through books of travel, Sir Northrop McMillan and his wife, both American born, whose Juja ranch is famous. We did not get out to the ranch; the McMillans' Nairobi home is a most attractive, delightful, gray stone place, covered with the blue blossoming plumbago, and Lady McMillan's racing stable houses about fifty lovely thoroughbreds.

We met, too, these well-known outfitters, Messrs. Newland and Tarlton, whom other expeditions usually meet on entering the country; we gazed upon the trophies of that Duc d'Orleans who had been on the *Kenilworth Castle* with us, going with his physician after big game; we heard the gossip of all recent expeditions. We met Major Dugmore, whose African pictures are so well known, just starting with the expedition of Mr. Harris of Detroit, a young graduate of Yale, anxious to obtain beautiful pictures of African animals.

A party of us motored to see a dance of the Kikuyus. Although they live the closest to the encroaching civilization, these men wear less than any savages we encountered in the interior; in many cases a string of beads and a literal fig leaf sufficed, and at times the fig leaf was omitted. The emphasis was entirely upon decoration. They were painted to the saturation point. Their hair glistened with ruddy ochre, their bodies were vivid with white and scarlet and lemon yellow and horrifying pea green. Each man had worked out a personal scheme of decoration; each had his intricate facial designs, vivid half moons and polka dots, and each body was zoned and ornamented according to the personal taste of the designer. They made a mad picture as they stamped and sang and surged back and forth, now in brilliant sunlight, now in the shades of the vast spreading sacred tree of two hundred years veneration.

Even the onlookers were smeared with the burnt sienna-colored ochre of the region, the red earth mixed with castor oil smeared on hair and skin and goat skins. Only the chief, shrouded in a scarlet blanket, disdained the paint. He took a fancy to Alice and led her out for the dance to go on about them. Alice was frankly tired of chiefs but she was intrigued by the Kikuyu's way of decorating his ear—by making an aperture in the lobe, and then distending it to such surprising size that a tin can is sometimes worn as an insertion.

On our way back we stopped for tea at the Newlands', where there were two dear little girls—Elaine and Margaret. Margaret's advent had been known to us twelve years before through John McCutcheon's cartoon of welcome to her, reproduced in his *In Africa.*

Another character of the East Coast we met in Nairobi was Cherry Kearton, whose African pictures have been often shown in America. He told us a magnificent story about lions. There were six lions and a tree. Mr. Kearton and a friend were in the tree. As

RIPON FALLS—THE BIRTH OF THE NILE

[page 250]

ALICE AT A CEREMONIAL DANCE OF THE KIKUYUS

[page 250]

NATIVES SAWING IN NAIROBI

[page 253]

A RICKSHAW AT MOMBASA

[page 253]

that story developed, in its strength and simplicity, I found myself believing in it, believing utterly. There was no reason why that story should not have happened to Cherry Kearton—he had been long enough in the country to have almost anything happen. I accepted it. I accepted all six of the lions. And then he added a detail. He spoke of seeing the light of their eyes—six pairs of gleaming eyes in the dark.

Now there is a school of romance that encourages belief in the gleam of a cat's eyes at night, but the only time that a cat's eyes gleam at night is when there is a light directly reflected by them. If there is a light in the room behind you and you look under the bed in a dark room your cat's eyes under the bed may gleam back at you. And if you are up a tree and have six lions around you and you flash a light down into the darkness, it may happen to be reflected from the eyes of a lion directly in front of that light; the other five pairs of eyes are gleamless. And if you are up a tree and you flash no light you get no gleam from any lion.

I do not say that this discredits the complete drama of Mr. Kearton's story. He may, like many another artist, have gilded his original lily. He did not think we knew the truth about a cat's eyes at night. But he knew it. . . . And somehow I find myself not believing in all six of the lions. Say about one and a half.

But we did believe in his tame chimpanzee, for he brought him to the hotel and the chimpanzee and Alice promptly cuddled up together.

There was another and grimmer lion story I heard from an old surveyor whose business led him among the native villages where he was on very friendly terms. One night he was being entertained in the chief's hut, where he slept with the chief who was remarkably fat, and two of his sons. In the middle of the night something, some pricking prescience of danger, made the white man stir. The tiny aperture of a door, the only entrance to the hut, framing a light square of sky, was suddenly blotted out. The next instant there was a scream, and then a growl and cracking of bones. A lion had entered and struck down one of the little black boys.

It was a horrible situation. And the white man in order to show his trust had left his gun in another hut with his other belongings. . . . Nor had the chief a knife. The three of them lay there quaking and silent, not daring to stir, while the lion munched down the chief's son, leisurely, hour after hour, it seemed. The beast was between them and the door; sometimes they could see his head silhouetted against it. Finally, gorged and replete, he rolled over against it, by the body of the boy and went off to sleep. He slept there all day. And those three lay there unstirring, afraid to move a finger. They could not dig themselves out through that bricklike floor; they could not break the reeds of the hut. The one thought in the surveyor's mind was that the lion would surely get thirsty and go out for a drink—he kept telling himself that over and over again. He would lie up all day and go out and drink at evening.

Evening came—the lion began to stretch and stir. . . . Then came a sound like a cat lapping milk—blood

—the boy's blood was his drink—then a sniff. It occurred to the surveyor that he might be hungry again, but there must surely be enough of the lad left for a second meal. He made horrible calculations. Then the lion began to walk silently about that hut. The door was visible again.

They made a dash for it, but fat as he was the native chief reached the door before the boy or the white man reached it and made a plunge and stuck. The door was edged with sapling uprights, through which it was always a squeeze to go in, and somehow in this crucial moment he stuck. He hadn't an instant's grace; the lion was after him. What followed is not pretty. The lion simply ate him, gradually and completely, while the boy and the surveyor, utterly helpless, lay there sick with terror. The only merit in the situation, and the surveyor admitted that he was callous with the thing by then, was that the lion was headed outward and was eating himself finally out of the hut. But, alas, the dimensions of the chief, that had caused him to stick, were no smaller in the lion. As the gorged beast pressed forward, he, too, felt himself caught; he pushed —the saplings gripped him, he could go neither forward nor back. He was caught, trapped with his swollen avoirdupois. And there he was stuck until he should starve down to thin dimensions again—with the surveyor and the remaining boy bottled up behind him.

The surveyor stopped. "And he did starve down?" I demanded at that point.

"Ultimately," said the surveyor. "But it took days."

I thought of the two imprisoned there in that horrible

hut. "But you—what did you live on?" I thought to ask.

"Oh,—*that,*" said the surveyor. "I lived on the boy."

From Nairobi the tourist trail runs down to the sea, three hundred and thirty miles to Mombasa and a descent of five thousand feet. Past Kiu and Tsavo and Simba it runs, names that are starred with anecdotal interest in the tourist's lexicon, across the wide plains, Kapiti and Athi, through the famous preserves where herds of wild game ramble past the rolling car windows.

This is the tourist's usual introduction to Africa coming in from the East Coast, but we found the trail as interesting on leaving the country as if we were seeing Africa for the first time—in fact there is a poignancy to last times that made every crooked little thorn tree an object of tender affection.

In these sun-burned reaches we saw none of the cob and topi that covered the Ruindi; in fact the topi is so rare in Kenya Colony that the hunter is limited to two, but we saw kongoni and waterbuck and gazelles, the large and graceful Grants with its splendid sweep of horns and the frolicking little Tommy, or Thompsons. with his black and white striped sides and his ever flirting tail. More novel and interesting than these to us were the zebra, of which we saw fifty in one herd, and the giraffe.

Our first glimpse of giraffe was of a procession of them rocking along against the sky line for all the world like little mechanical toys; then we passed a long line very close to the train, twenty-four of them strung

Giant Baobab Tree at Mombasa

[page 254]

Good-by to Africa

[page 254]

MEMORIES

[page 255]

along the way with their amusing heads hung up in the sky as if suspended from nothing at all, staring down earnestly upon us. Then one by one each turned and galloped away. The gait of the giraffe is the final joke of its appearance, a trot with the first two legs and a rocking gallop with the last. This sounds as if it were intended for a joke like the gait itself, but its truth is vouched for by no less a naturalist than Mr. Akeley, who affirms that he is its discoverer, and after the constituent elements are pointed out, you can easily distinguish the paces for yourself.

Mombasa was a blaze of white coral—a little island three miles by two, glittering like a frosted birthday cake with that white rock of which its foundations and its buildings are made. It was as tropical as a dream of the South Seas, tufted with the cocoanut palms and baobabs,—giants of trees with huge bottle-shaped trunks—spangled with scarlet bloom and brilliant with sunshine from a fervid blue sky.

It was a pocket of contrasts. It rattled with rickshaws and motors. Green golf links stretched out to crumbling ruins of old Portuguese forts. On a street corner an English girl and a black *bibi* in a yard of calico and a veil-shrouded woman of the East. . . . Tennis and tea shops . . . tourists . . . tailors sewing cross-legged and yellow Indian babies with pearls dripping from nose and ears minded by patient black boys . . . a swimming club and a lone row of cocoanut palms and a white beach where the Indian ocean rolls lazily in, and painted dhows rock in the tide and cattle bellow in their holds. . . .

Out across that Indian Ocean ran the trail for Home. Up from Mombasa to Guardafui, through the Gulf to Aden on its barren rocks, over to the sunbaked camel squares of Jibouti, the mouth of Abyssinia, up through the Red Sea and the ribbon of the Suez Canal, through the Bitter Lakes, past Port Said, where the East and West meet in unholy traffic, and across the Mediterranean to Marseilles, and a racketing train through France to a blowy channel. Then England. Then America. Then Home.

Ahead of us ran the trail. Behind us was the old Africa that we knew and had learned to love, the Africa that had been going on from the beginning of time, the wild and lovely land untouched of man. Lost to us were the vast spaces, the splendor of waters, the mad glory of volcanoes, the fairy isles of hidden lakes, the enchantment of cloud-wrapped heights; lost the solitude and the beauty and the freedom that make our civilization seem a prison and a market place.

Gone, too, was the kindly security of the wilderness, the open tent flap, the money box left carelessly without, the unguarded ways. I was sorry to take the Littlest Explorer back to charging motors and barred doors to face with her again the perils of civilization.

Africa had been so worth while. And no after-years can take the memories from us—the black outline of a lion against the moonlight . . . the sheen of the golden crested crane . . . women with water jars crossing the opal sands of Tanganyika . . . the long, superb line of Mikeno cutting the tropic sky . . . a glade in a fairy forest and a great gorilla in the sun.

CHAPTER XX

LISTS AND EQUIPMENT, ETC.

EQUIPMENT is of interest only to those contemplating an African trip and then it becomes of feverish importance. Every one's experience is different and sometimes contradictory, but the accumulation of individual detail often throws light on a subject. For the benefit of those who want to know, I am adding some lists of what we took with us and trying to answer the questions of what we really wore and ate and needed on such an expedition.

Our experience was unlike that of the usual traveler who enters the country by the East Coast and engages the services of such a firm as Newland, Tarlton and Company, Limited, from whom he can obtain any part or all of his equipment and the services of porters, gun boys, tent boys, and cooks for the entire time that he is in the field. In the past Mr. Akeley had got his things from England and engaged his men from Newland, Tarlton. Colonel Roosevelt had outfitted in America and got his men from Newland, Tarlton. I met men who outfitted personally at the Nairobi stores and engaged their men themselves.

For our safari we brought everything from America and England, picked up boys as we got into the Congo and carried them through with us, and obtained porters at different posts and villages. We changed porters eight times. We took our goods up the long way north from the Cape as excess baggage instead of freight, in order to have it travel on the same trains with us. Mr. Akeley was already provided with some of his bags and nets so the orders in the ensuing lists are often for odd numbers. The following supplies were all obtained from Benjamin Edgington, 1 Duke Street, London Bridge.

Camp Equipment

1 "Whymper" tent complete.
3 double roof ridge tents, 10 by 8—four feet walls, in valises.
1 extra fly, with uprights, ropes, complete.
3 ground sheets from heavy green rot-proof canvas 11 by 9.
4 mosquito nets, fine mesh, for half tent.
3 circular canvas baths.
6 holdall bags with bars and padlocks.
6 green round bottomed bags fitted with eyelets and cords, 43 by 30.
5 enamel wash basins.
6 "Uganda" waterbottles, 8 pints.
2 Machilla hammocks, green canvas, with awnings, double strength.
12 Bath towels.
18 Face towels.
6 Ibea folding chairs.
6 best quality deck chairs.
6 air-tight boxes.
6 small green hair pillows.
4 candle lanterns.

Our cots were the Gold Medal, Racine, Wisconsin, extra size. Our dining tables were a special design of

Mr. Akeley's made for us by a friend, Mr. Clarence Dewey, in New York. Our kitchen ware, tools, such as hammer and hatchet and rope—it is essential to have a good deal of rope—were bought at Elizabethville where we bought our bicycles. We were unable to obtain the desired petrol lamps in the Congo so we were obliged to depend on our candle lanterns. It would be wise to come provided with a good light. We had both the ordinary type of flash light with extra batteries and the type that creates its power by constant squeezing and both are desirable.

Our food supplies were all obtained through Edgington's, and came put up in "chop boxes," wooden boxes with lock and key, packed not to exceed a porter's weight of sixty pounds, and marked for identification. These supplies were the following:

TWENTY CASES, EACH CONTAINING

2 tins cheese, Cheddar, Gruyere, Camembert.
1 pound tin Ceylon tea.
4 pound tin granulated sugar.
4 tins sardines in oil.
2 1-pound tins rolled ox tongue.
3 tins Underwood deviled ham.
5 tins jam, assorted, no currant.
2 tins Dundee marmalade.
4 one half-pound tins Danish butter.
5 one half-pound tins beef dripping.
5 half-pound tins Ideal milk.
2 number 2 tins small captain biscuits.
4 tins Heinz baked beans and tomato.
1 small tin Cerebos salt.
2 one-pound tins plain chocolate.

1 one-and-a-half pound tin Scotch oatmeal.
1 one-half pound tin baking powder.
1 box about two-and-a-half pound primrose laundry soap.

TWENTY CASES, EACH CONTAINING

2 tins Heinz baked beans and tomato.
2 tins smoked sardines in oil.
2 tins smoked brisling in oil.
2 tins camp pie.
5 tins jam assorted, no currant.
2 tins Dundee marmalade.
5 one-half pound tins Danish butter.
5 one-half pound tins dripping.
5 half-pound tins Ideal milk.
2 tins cheese, Cheddar, Gruyere, Camembert.
1 one-pound tin Ceylon tea.
1 three-quarter-pound tin ground pure coffee.
1 four-pound tin granulated sugar.
1 one-quarter-pound tin pure cocoa.
4 No. 1 tin camp biscuits, plain, various.
1 small tin Cerebos salt.
1 one-and-a-half-pound tin Scotch oatmeal.
2 two-pound tin prunes.
1 one-ounce castor ground white pepper.
2 sponge cloths.
12 quire kitchen paper.
2 one-pound tins plain chocolate.
1 bar primrose laundry soap.

EIGHT CASES, EACH CONTAINING

6 seven-pound tins flour, special export.

TWO CASES, EACH CONTAINING

15 bottles Montserat lime juice.

LISTS AND EQUIPMENT, ETC.

TWO CASES, EACH CONTAINING

30 pounds Wiltshire bacon.
canvas and salt.

FIVE CASES CONTAINING

4 tins whole Edam cheese.
20 2-ounce tins Bovril.
10 2-pound tins Sultana raisins.
10 1-pound tins macaroni.
30 4-ounce tins Underwood's deviled ham.
20 bars carbolic soap.
10 bottles Enos fruit salt.
10 1-pound tins Christmas pudding.
6 one half-pound tins curry powder.
10 one half-pound tins yellow dubbin.
20 bottles of Chutney (10 Bengal, 10 Major Grey).
4 pound tins veterinary vaseline.
6 1-pound tins castor sugar.
6 Knight's patent tin openers.
24 tins asparagus tips.
24 tins Black Leicester mushrooms.
6 large bottles Pond's Extract.
12 10-yard spools Z.O. surgeon's tape, 1 in. wide.
4 bottles Worcestershire sauce.
6 tins best mustard.
12 dish towels.

THREE CASES CONTAINING CANDLES, PLASTER OF PARIS AND THE FOLLOWING TABLEWARE:

12 white enamel plates, light weight.
12 white enamel dinner plates.
3 white enamel vegetable dishes, medium size.
6 one-pint cups (mugs).
6 white enamel oatmeal dishes.
6 cups and saucers.
12 extra saucers.

Of that collection of food, only two boxes were untouched when we came out of Uganda. We were well pleased with the assortment and the Danish butter and Ideal milk were exceptionally good. Some of us ate our tinned butter and milk by preference, even when we could have fresh milk and butter at the posts. Another time we should want more coffee, and add oil and vinegar to the lists; we should want more of the Edam and less of the tinned cheese, less drippings and beans and omit the mushrooms altogether. Going into the Congo again it would be well to bring in some tinned fruits; in Kenya Colony on the East Coast, these are usually obtainable from Indian venders. In the Congo we were always able to obtain chickens and eggs from the natives and, almost invariably, bananas; our table from time to time knew green corn and fresh tomatoes, onions, leeks, cabbages, peas, beans, marrows, potatoes, both white and sweet, pineapples, oranges, papayas, mangoes, limes, strawberries, and gooseberries.

Medicines

Our medicine kit was a Burroughs and Wellcome medicine case, supplemented with ample oxide adhesive plaster, bandages and hypodermic syringe in case of wounds and blood poison. Iodine, quinine, and adhesive tape are essentials. A dental kit is advisable, made up of the simplest remedies and amalgam, put up by one's own dentist—and it is an excellent idea to have one of the party spend an hour with a dentist imbibing first aid to the aching.

LISTS AND EQUIPMENT, ETC.

Cameras

Mr. Akeley's equipment was elaborate, and included two Akeley motion picture cameras, one being "The Gorilla," a stereoscopic camera, a Graflex camera, and a dark room and equipment for developing negatives and making motion picture film tests in the field. Mr. Bradley used a Graflex plate camera of Mr. Akeley's, and Miss Miller and I had each a 3 A Kodak. Mine inexplicably collapsed in the middle of a film in Uganda. Miss Hall used a Brownie. Our films were brought from America done up in lead. We could have bought them at Elizabethville but no one ought to depend upon the chance of getting the right film at the last moment. Neither our plates nor films suffered any deterioration. The plates are extra trouble on safari but the results are well worth it. I can say that unqualifiedly, not having been the one to take the trouble! For quick snaps about camp a universal focus is a good thing to have at hand. We had four porters carrying plates and five porters carrying motion picture film.

Guns

Mr. Akeley and Mr. Bradley each carried a Jeffery's .475, two rifles which had already made one trip to Africa with Mr. Stevenson and Mr. McCutcheon. The .475 is a heavy double-barreled, cordite rifle, of tremendous stopping power, an invaluable thing for emergencies. The men endorsed it heartily. Any gun of that type is considered too hard in recoil for a woman. Mr. Bradley used an extra rubber pad on the gun stock and never felt any lameness although he fired as often as twelve times on some days.

Each one of us was equipped with a Springfield army rifle, with sport stock built to order and special Lyman sights. Mine was directed against so few animals that I can do little more than give it a good character for the times when I needed it—against lions. We supplemented this arsenal by the purchase of a 16 gauge shotgun at Elizabethville.

Clothing

My outfit for safari was modeled on that of the men. I had two khaki suits, coats and knickers with loose knees, an extra pair of riding breeches and a khaki skirt. In the field I used flannel shirts, but on the march I usually wore a white silk or heavy crepe blouse with the khaki coat and knickers. I never wore a spine pad; Alice wore hers only on two days. I had a belt, two leather pockets, a knife, a compass. I had three pairs of leather boots, two heavy, hobnailed, leather-lined and one of softer leather, and one light pair of rubber-soled canvas, which were very useful, especially on the bicycle. I wore cloth puttees, of which I had three pairs. I had eight pairs of Jaegar stockings, four heavy and four fine, two pairs of bicycle stockings, and two pairs of long bed stockings. I wore silk underwear except for the days on the gorilla mountains when I wore Jaegar wool. I had two suits of warm pajamas, a heavy sweater, a rain coat, a pair of mosquito boots, soft, leather topped boots for camp wear, a gray pith helmet, and the usual toilet articles, handkerchiefs, writing materials, etc.

Alice wore exactly what she wore in Wisconsin wilds—khaki knickers and middies, over cotton and silk underwear or sometimes over the lightest weight wool. Her stockings were the finest and softest wool; her boots were stout elkskin and ponyskin, and she wore canvas leggings.

The field outfit was simple. The lists that were really appalling were for the rest of the journey. We had to plan for both cold and heat. For the voyage over, one wants plenty of white sport things and evening clothes. Capetown was cold enough for coats at times, but generally a silk frock was warm enough. On the long trip north on trains and steamers, I found that a predominantly dark sport skirt and a dark blue silk sweater with white blouses were useful. All through the Congo I carried one air-tight packed with the evening dresses that I should need when I came out on the East Coast, and with white afternoon wear when we struck an official Boma. Seven dresses, two sports skirts, four blouses, and six pairs of slippers and shoes, and the various "undies" lived in that 10 inch by 14 by 32 air-tight! For the survival of the fittest in evening gowns I recommend lace ones and heavily beaded georgette—the beads weight down the wrinkles. But this is becoming a feminine chapter.

"Dangers"

Many questions have reached me about the dangers of an African trip. I can only say that except for the greater distance from physicians in case of a broken arm or leg an African camping trip doesn't seem to me more dangerous than one in California or the Rocky Mountains. There is danger from sun, but not if you wear your helmet. There is a chance of fever, but ordinarily nothing that quinine won't take care of, and the immunity from colds and influenzas and the contagions of civilization is well worth having. I saw only two snakes in six months in Africa; one was dead, and one was an infant. Neither was harmful. We had to go out and hunt hard for all the dangers that we had—elephants and lions and gorillas. These can easily be avoided by any one who wants to see the country and is not anxious for combat. The natives along the line of march we found peaceful, and I do not believe that there would have been much chance of trouble even among the insurrectionists for they are not in a state of violent insurrection but merely hold aloof from the white man's rule.

With time and money and health any one can see Africa. To see the hidden Africa, the wild Africa, that swiftly vanishing savage land, takes more time and a little more money, a passion for exercise and an enthusiastic digestion. Our experience certainly showed that with these and with good care, not only men but women and a little child might go safely into the very heart of the Continent.

INDEX

Albertville, 48.
Antelope (gazelles), 254.
  Karisimbi, 122.
  Kenya Colony, 199, 248.
  Pet, 88.
  Ruindi Plains, 177, 187.
  Situtunga, 240, 241.
Ant hills, 42.
Ants, safari, 64, 65.

Baboons, 200, 201.
Baganda, 230.
Bamboos, 108, 223.
Bark cloth, 201, 202.
Barns, T. Alexander, 81.
Baron Dhanis, 51.
Baskets, 92.
Batiga, 230.
Batwa, 136, 137, 138, 139, 140.
Belgian Administration, 217, 218.
Bicycles, 59, 60, 173, 188, 190.
Big Game, dangers of, 168, 169.
  on Ruindi, 199.
Birds, 238, 239.
Borassus palms, 44.
Boys, characteristics of, 165, 166, 228, 243, 244.
  costumes of, 143.
  discharge of, 144.
  payment of, 41, 162.
  relations to, 166, 167.
Bubonde, 102.
Buffalo, dangers of, 168, 169, 206.
  hunting, 208, 209, 210, 211.
  on gorilla mountains, 114.
  on Ruindi, 206, 207.
Bukavu, 73.
Bulawyo, 32.
Bunyoni, Lake, 223, 224, 225.
Burton, 51.

Camp life, 61, 62, 63, 157.
Cannabalism, possible case of, 70.
  former, 74.
Capetown, 25, 26, 27.
Cecil Rhodes' grave, 32, 33.
Changugu, 72.
Chaninagongo, Mt., 79, 95, 104.
Chiefs, greetings of, 61, 221.
  ivory of, 84.
  presents of, 60.
  sociability of, 82, 83.
  wives of, 62, 84.
Chimpanzees, 104, 133, 251.
Christmas, 213, 214, 215.
Cicatrization marks, 45.
Cob, 177.
Cold, 119, 120, 156.
Cooks, 157, 158, 159, 160, 163, 164.
Crocodiles, 46, 47, 48, 72.

INDEX

Dances, native, 74, 138, 141, 249, 250.
Daudi, King of Uganda, 233, 234.
Deriddar, Dr., 79.

Elephants, dangers of, 168, 169.
  hunting, 59, 65, 66, 200, 202, 203, 204, 205.
  presence of, 44, 108, 205.
Elizabethville, 38, 39.
Entebbe, 238.
Equipment, cameras, 263.
  camp, 258, 259, 261.
  clothing, 264, 265.
  food, 259, 260, 262.
  guns, 263, 264.
  medicine, 262.

Fever, 77, 79, 242.
Filed teeth, 70, 74.
Fire sticks, 202.
Flies, 61.
Food, in the Congo, 46, 62, 80, 81, 141, 142, 161.
Forests, on gorilla mountains, 114.
Freetown, 14, 16, 17.

Gavial, 48.
Gazelle, 254.
Giraffe, 254, 255.
Goma, 78.
Gorilla, band of, 123, 124, 125, 133.
  beds, 128, 133.
  difficulties of seeing, 81, 109.
  Du Chaillu's description of, 4, 105.
  equatorial, 104.

Gorilla (*Cont.*)
  food of, 131.
  forests of, 114.
  hunting, 106, 107, 108, 109.
  killed by Mr. Bradley, 115, 116, 117.
  lack of information about, 4.
  meat, 121.
  necessity for preservation of, 132.
  Prince of Sweden's expedition for, 104.
  raids by, at Katana, 75.
  raids by, at Walikali, 76.
  shot by Mr. Barns, 99, 100.
  tracks, 106.
Groote Schur, 28.
Grotzen, Count, 71.

Hail, 147.
Hairdressing, native, 45.
Hippopotamus, 45, 85, 174, 248.
Housekeeping in the Congo, 88.

Ivory, 89, 90.

Jackals, 195.
Jiggers, 84, 85, 112.

Kabale, 226, 227.
Kabalo, 48.
Kaffir-Boom, 63.
Kampala, 231, 232, 233, 242.
Karisimbi, Mt., 78, 114, 145.
Katana, 75.
Kavirondo crane, 88, 89, 239.
Kibati, 101.
Kigoma, 51.
Kikuyus, 249, 250.
Kimberley, 30.

INDEX

Kissenyi, 78, 87, 89, 92.
Kitchen, three-stone, 161, 162.
Kivu, Lake, description of, 72.
  discovery of, 71.
  climate of, 85, 86, 87.
Kongoni, 248, 254.
Koto-Koto, 60.

Lava, fields, 147.
  flow of, at Kivu, 94, 95, 96.
Leopards, 169, 170, 173, 175.
Lions, dangers of, 169.
  daylight hunting of, 177, 178, 179, 180, 181, 182, 183.
  grunting of, 185, 188.
  hunted by, 188, 189, 190, 191, 192.
  night hunting of, 185, 186, 193, 194, 195, 196, 197, 198.
  stories of, 250, 251, 252, 253, 254.
Livingstone, 34, 51, 71.
Lualaba River, 42, 44, 45.
Lulenga Mission, 103, 104, 134, 171.

Madeira, 12.
Mafeking, 31.
Mai ja Moto, 175, 217.
Mail, to Congo, 87.
Marabou, 205, 239.
Mbarara, 231.
M'fumbiro Mountains, 95, 145, 146.
Mikeno, Mt., 78, 106, 107, 108, 111, 112, 145.
Missions. *See* White Fathers.
Mombasa, 255.
Monkeys, 219.
Mosquitoes, 61, 77.
Motors, 231, 238.
Mountains of the Moon, 95.
Musinga, king of Ruanda, 61, 91.
Mutesa, tomb of, 234, 235, 236, 237.

Nairobi, 248, 249.
Naivasha, Lake, 248.
Nettles, 122.
Nile, sources of, 95.
Nyamlagira, Mt., 145.
  ascent of, 146, 147, 148, 149.
  camp on, 152, 153.
  cave in, 152.
  crater of, 149, 151, 152, 153, 154, 155.
  eruptions of, 146, 154, 155.
Nyunde, 80, 93, 94.

Otters, 225, 247.

Papyrus, 223.
Papaya, 62.
Polygamy, 229.
Pombe, 53, 160, 161.
Porters, clothing of, 62.
  engaging of, 58.
  fighting amongst, 70.
  food of, 62.
  status of, 58.
  wages of, 58.
Posho, 62.
Provisioning. *See* Equipment.
Pygmies, 136, 137, 138, 139, 140.

Quinine, 77, 80.

Rains, 44, 86, 150, 172.
Reed buck, 177.

INDEX

Rhinoceros, dangers of, 168, 169.
none on Ruindi, 170.
Rift Valley, Central African, 95, 145.
Ruanda, king of, 61.
kingdom of, 78.
inhabitants of, 79.
Ruchuru, 171, 216, 220.
Ruchuru River, 173, 174.
Ruindi, plains of, 175, 177, 199, 212.
Ruindi River, 175, 199, 201.
Rusisi Mountains, 71.
Rusisi River, 63.

Sake, 76.
Secretary bird, 42.
Serval cats, 195.
Sese Islands, 240.
Sharpe, Sir Alfred, 96.
Silver trees, 28.
Situtunga, 240, 241.
Sleeping sickness, 240, 241.
Smuts, General Jan, 16, 19, 20, 21, 22, 23.
Speke, 51, 237, 238.
Spirillum fever, 79.
Stanley, 42, 51.
Swahili, 82.

Tanganyika, Lake, 48, 49, 50, 51.
Thanksgiving, 141, 142, 143.
Tick fever, 61.
Topi, 177.
Tsetse fly, 59, 240, 241.

Uganda, boundary of, 220.
march through, 221, 222, 227, 228.
water, scarcity of, in, 221, 228.
Ujiji, 51.
Usumbura, 51, 52, 53, 54.

Victoria Falls, 33, 34, 35.
Victoria Nyanza, 238, 247.
Visoke, 78, 95.
Volcanoes, 79, 95, 145, 146.

Wahunde, 119.
Wahuti, 90.
Walikali, 76.
Waregga, 73, 74.
Wania-Bongo, 73.
Water, Scarcity of, 228.
Watussi, 90.
White Fathers, at Katana, 75.
at Lulenga, 103, 104, 136, 143.
at Nyunde, 80, 93, 94.
White Sisters, 94, 103.

Yubile, 147.

Zambezi, 35.
Zebra, 241.